Contents

AF718757

PLANNING

Codes and Regulations 3
Site Selection
Housing 4

HOUSING

Barn Styles 5
Open Front Buildings 8
Barn Layouts 9
Barn Construction 12
Manure Handling 19
Fire Safety
Rodent Control 20
Windbreak Planning

VENTILATION

Open-Front Buildings 22
Cold Enclosed Housing 24
Warm Enclosed Housing
Insulation
Vapor Barrier 25
Supplemental Heat 26
Ventilation Systems
Fans 27

CONSTRUCTION

Tack Room Equipment 29
Fences 31
Gates 33
Automatic Gate 34
Hinges, Latches 36
Windbreak Fences 37
Jumps 38
Dutch Doors, Mineral Feeder 40
Cupolas 41
Loading Chutes 42
Show Box, Breeding Stall 43

MATERIALS

Wood 44
Metal, Masonry 45
Concrete 46
Paint 48
Fasteners 50
Glue 52

BUILDING CONSTRUCTION 53

PLANS LIST 54

REFERENCE DATA 58

INTRODUCTION

Stable management is among the important activities of a good horseman, and the buildings and facilities, and their arrangement, very much affect this activity. Carefully planned, well-designed facilities make horse care easier and add to personal satisfaction and riding enjoyment.

In planning or acquiring a facility for horses, the essentials are knowing what is needed, how to get it, and keeping within the budget.

Two of the foremost considerations in planning a facility for horses are health and safety—not only of the animals, but also of the people who either come in contact with the animals and facilities or who live nearby. Other important considerations include sound construction, labor-saving conveniences, a building style in harmony with the surroundings, and a planning procedure that lists needs and wants in order of importance and that places necessities above mere desires.

The same kind of sound basic planning is required when selecting and equipping a barn to stable only one or two animals as for a larger number. Many styles and sizes of barns and other facilities are used to house and care for horses and to provide space for associated activities.

Whether you plan to build new, or to use an existing facility, go through the planning procedures for your entire layout. These include general management and financial decisions, site selection, site development, and building selection. Provide room for expansion and allow for specific details, such as water supply, manure disposal, feed storage, exercise space, loading and unloading access, and legal or other restrictions that may apply.

The size of your spread will substantially affect needs and costs and will modify the importance of different aspects suggested for planning consideration. But, with future expansion as a possibility, it's well to keep in mind all aspects, even though your immediate plans may be limited.

PLANNING

ESSENTIALS FOR PLANNING

Here is a listing of some planning elements to check, decide on, or determine before moving on to detailed planning or starting the project. The more you know in advance, the better you'll be able to plan and to avoid or cope with difficulties that might occur unexpectedly later.

- Look into building codes, zoning regulations, and any other local restrictions that may affect development, construction, or use of the facilities.
- Collect ideas on horse facilities from publications, tours, visits to existing layouts, and from experienced horsemen, professional managers, etc.
- Choose a style for the barn or other buildings that fits the site and is in harmony with the surroundings.
- Decide on the number of animals to be handled and on how they'll be housed and managed.
- Determine the total amount of barn space needed under roof to stable and care for the animals. Basic space needs are for stalls, traffic lanes, and storage of feed, tack, and equipment. But include, also, special-purpose features: wash rack, trailer storage, breeding area, show stable, riding arena, rest space, and other areas needed under the same roof.
- Figure the amount of space needed under roof for separate but associated buildings; for example, open shelters, riding arena, sales barn, training barn, equipment storage, exercise area, hay and feed storage, office and lounge, and other areas (including living quarters if needed).
- Determine the amount of space needed for open and fenced areas: lanes and roads, outside lots and corrals, exercise areas, parking space for trailers and other vehicles and equipment, special events, and distance between buildings for pleasing appearance, fire protection, and future expansion.
- Compare alternative housing arrangements and management methods and select those most suited to your situation.
- Investigate methods and plan for inoffensive manure storage and disposal.
- Find out what fire protection is available, and modify plans accordingly for added protection or fire control if necessary.
- Review your budget allocation, placing priorities on necessities, conveniences, over-all desires, and frills in that order.

Once you've done these things, it's time to determine the size needed for the whole layout. Exclude family housing, large pastures, and crop and other land not directly associated with the stabling and facilities complex.

The size of the total layout is determined by the land area covered by the necessary buildings and the directly related open and fenced areas. This provides the basis for establishing the total land requirement. Large new and expanding horse farms may require sizable acreages, but only a small acreage may be needed for the building site and an exercise area for a new small spread. (If an existing building is used for housing only, remember that space for an exercise area may be needed.)

The information gained in determining space requirements for the present, as well as possible future expansion, can be used for developing completely new layouts. Pertinent parts also can be selected and used for expanding present layouts or starting ones that incorporate existing buildings.

CODES AND REGULATIONS

Early in the planning stage, check to find out what effects various regulations and restrictions may have on the proposed project. Society places certain limitations on ownership, development, and use of property.

Zoning regulations, building codes, and sanitary regulations have come along with the encroachment of urban areas into what once were unrestricted rural areas. Many of these regulations now are on a county, state, or national basis, and are intended to regulate property uses considered detrimental or adverse to public welfare. Generally, building codes set construction standards, zoning prohibits the use of property for specific purposes, and sanitary regulations have to do with public health, related pollution, and pest control.

Other restrictions that may limit the use of property include deed restrictions, easements, and covenants between property owners. Neighbors may become concerned about nuisances, whether there are specific regulations or not. So, check neighbors' reactions (and anticipate future neighbors' reactions) to your stabling of horses—especially if you live in an urbanized or developing area; ownership of urbanized property can change rapidly.

SITE SELECTION

The building site should be well drained, accessible, and have a slope of about 5'/100' away from the building in all directions to assure good surface drainage. Consider the grading and filling that will be needed for a well-drained site; plan on using only clean soil, sand, gravel, or crushed rock for fill.

Topography of the site, particularly, affects the cost of site development, and site preparation should be completed before building construction is started. A nearly level site usually involves the least cost. Sites on steep slopes or rocky terrain, or sites requiring considerable fill, are costly to develop and may make compromises necessary.

Among factors influencing selection of a building site are:

- **Size of operation**—In addition to room for the planned buildings, the site should provide space for other planned facilities and areas. It should also provide for future building and paddock expansion, and for good traffic patterns for safe and convenient handling of animals, vehicles, equipment, materials, and snow removal.

• **Restrictions**—The zoning regulations, building codes, sanitary regulations, deed restrictions, easements, and covenants mentioned earlier may affect or limit the development and use of property at a specific location.

• **Topography**—Surface characteristics of the land (such as direction and steepness of slope, gulleys, ravines, streams, swamps, knolls, hills, and rocks) affect site selection, development, convenience, and use. Boulders and rock outcrops or formations near the surface indicate that costly rock excavation may be required to develop the site and to install foundations and water and drainage lines.

Avoid sites that have serious drainage problems (such as steep slopes that concentrate surface runoff in the building area and wet areas caused by critical ground-water conditions) unless the problems can be completely eliminated. Avoid sites near streams that flood, and stay well above the highest flooding.

• **Water supply**—An adequate all-year water supply of the quality needed must be available at the site, either from a public water system or from convenient ground-water and surface water developments.

• **Nuisances**—Distracting noise, offensive odors, and heavy traffic may affect site selection. Consider existing nuisances that may affect you and your use and enjoyment of the site and also ones that you may create that might affect neighbors.

• **Windbreaks**—The natural protection from the elements provided by wooded areas, knolls, hills, and ravines is a consideration in site selection.

• **Existing buildings**—Existing buildings may be a determining factor in site selection, but only if their size, location, physical condition, and inside arrangement fit into the over-all plan for housing facilities.

• **Services**—The availability of electric service is essential.

• **Economy and aesthetics**—Appearance, personal taste, pride of ownership, and available funds may affect decisions in site selection.

• **Building layout**—Chances are that building layout will be a deciding factor in selecting the site, but the site chosen may affect both building layout and building style.

HOUSING

The horse barn, large or small, should be well planned, durable, and attractive. Its basic purpose is to provide an environment that protects the horses from temperature extremes, keeps them dry and out of the wind, eliminates drafts through the stables, provides fresh air in both winter and summer, and protects them from injury.

In cold uninsulated or insulated barns, fresh air usually is provided by natural air movement through wall openings and ridge ventilators. Wall openings may be small windows, wall panels, or slots under the eaves. In tight, warm barns, insulation, fans, and spaced air inlets are necessary. In cold climates, supplemental heat may be needed.

Provide ample space in the barn, based on established space requirements, for the well-being of the animals, the convenience and safety of the people caring for them, and the enjoyment of the rider.

Stables and other facilities have features peculiar to their functions. These features are incorporated in two basic building shells: open-front buildings for minimal shelter, and completely enclosed buildings for optimum shelter. The usable space in open-front or enclosed basic shells may be confined only to the ground-floor level. Depending on wall height and roof design, however, additional space overhead for hay, bedding, and feed storage may be provided in both open-front and enclosed buildings in a mow or partial loft.

HOUSING

BARN STYLES

Barn styles are determined by the distinctive shape of the roof. The three shapes most widely used are shed, gable, and offset gable. The once popular gambrel, gothic, and monitor roofs, while very much in evidence on old barns, are used only occasionally on new ones.

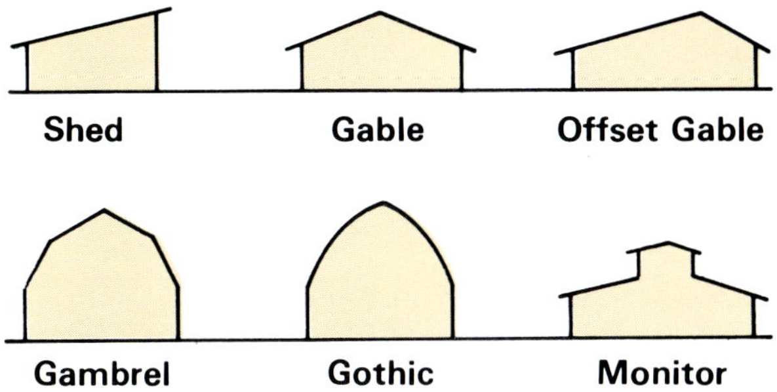

The roof style and the size and spacing of the roof structural members (rafters, beams, trusses, purlins, arches, supports, and braces) and the type of roofing used are determined by engineering design.

The trend is toward the use of one-floor structures with clear-span or post-and-beam supported roofs. With one-floor structures, costly overhead storage is eliminated or held to a minimum. The use of one floor also permits lighter, less costly, construction. Clear-span roofs also add flexibility to buildings by making them more adaptable to other uses.

Shed Roof

The shed roof is widely used on both open-front and enclosed permanent freestanding facilities, attached lean-to structures, and small movable buildings.

The shed is a single slope roof, a basic form for all sloped roofs. It is relatively low in cost, provides good headroom, and is simple to build and insulate. It is adaptable to natural lighting with spaced translucent roof panels.

Most shed roofs employ a low pitch (slope) to keep the high side of the roof as low as possible unless room for overhead storage is desired. The minimum roof pitch (slope) is determined by the manufacturers' recommendations for the kind of roofing selected.

Framing

Common rafters, beams, and trusses are used as framing members for shed roofs.

The most common spans for wood rafters range from 10'-16'. In wider buildings, interior posts and beams are used in addition to the walls to support the rafters. Stall and pen partitions influence interior post location.

To obtain a clear-span shed roof, one-slope trusses and long-span beams with purlins are often used. Clear-spans up to 48' are not unusual.

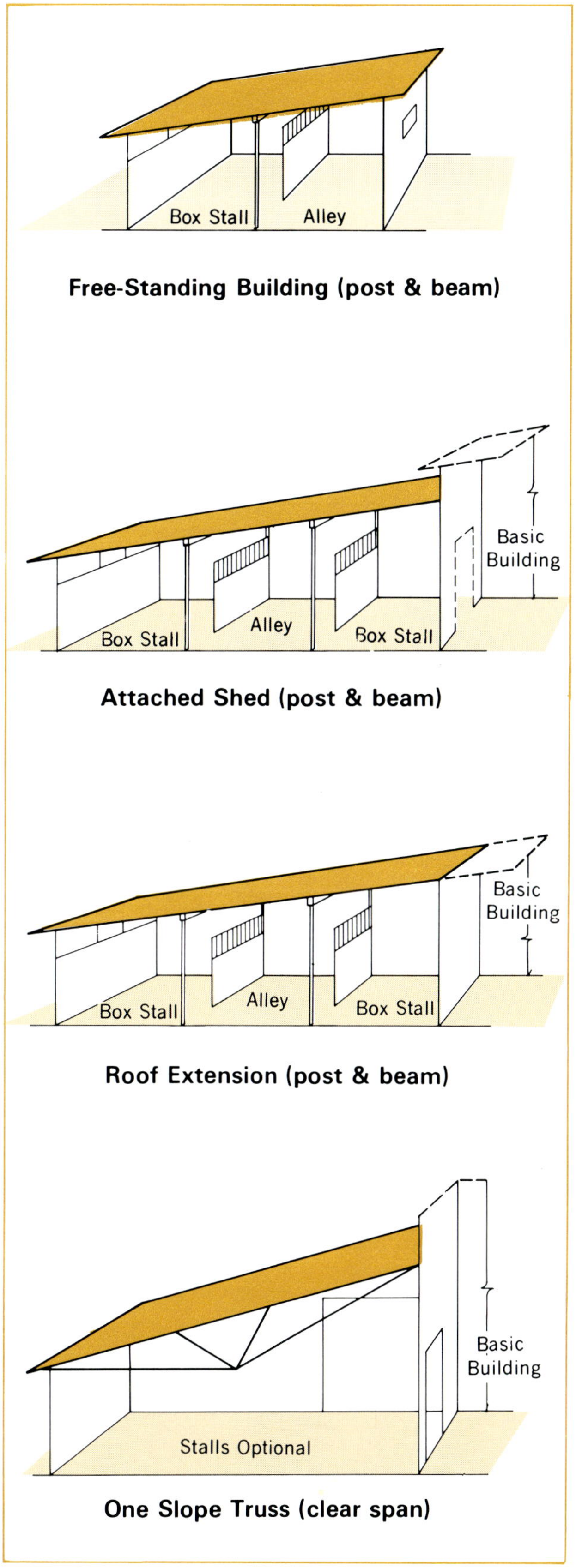

Shed roof options

Gable roof options

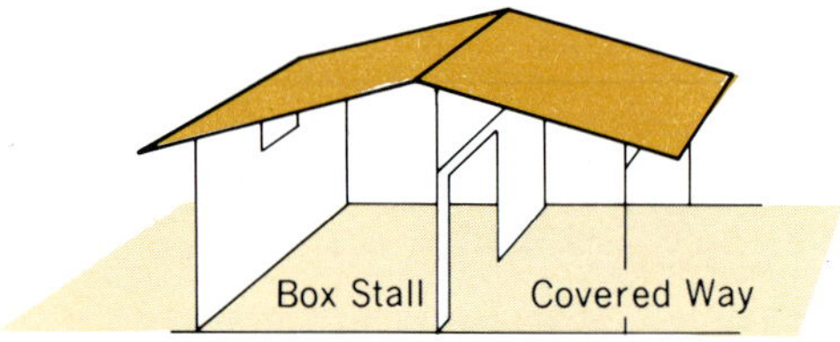

One Row Of Stalls Serviced From Covered Way
(post-beam-wall)

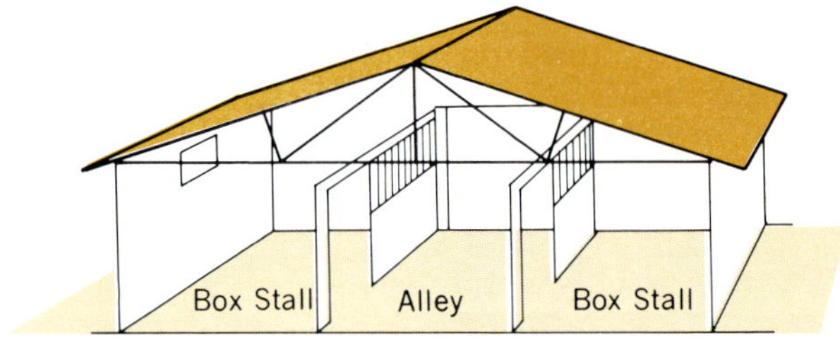

Two Rows Of Stalls Serviced From Center Alley
(clear span truss)

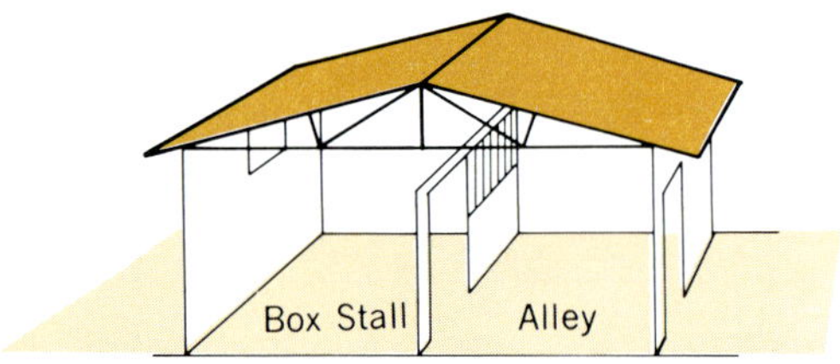

One Row Of Stalls Serviced From Enclosed Alley
(clear span truss)

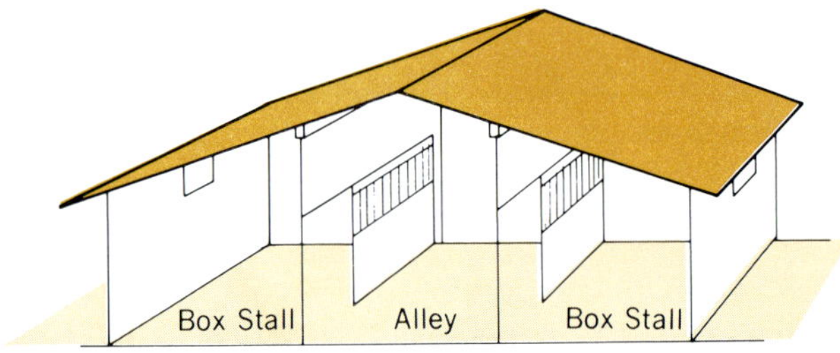

Two Rows Of Stalls Serviced From Center Alley
(post & beam)

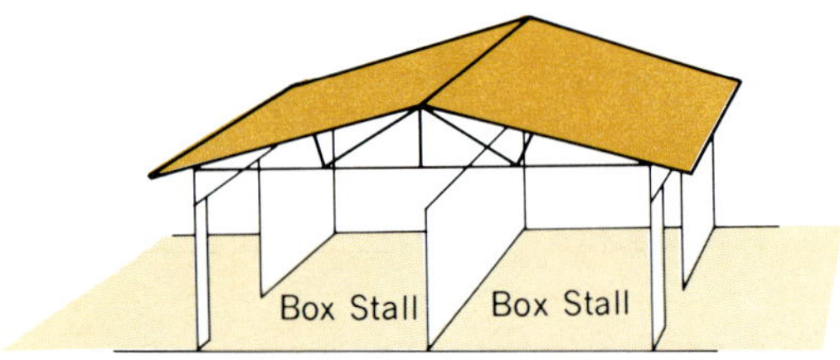

Two Rows Of Stalls Serviced From Outside
(clear span truss)

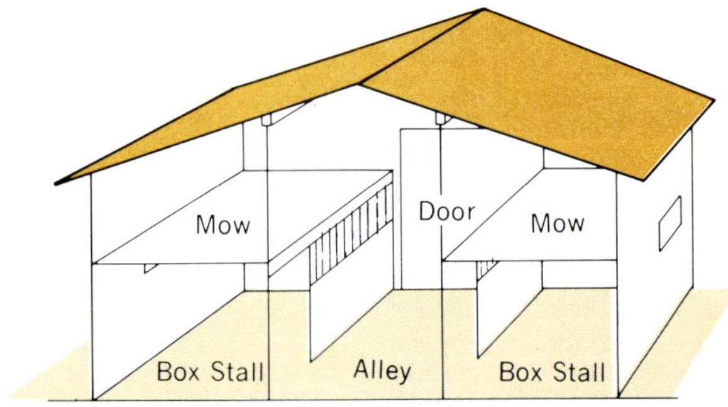

Two Rows Of Stalls Serviced From Center Alley
Mow Over Stalls (post & beam)

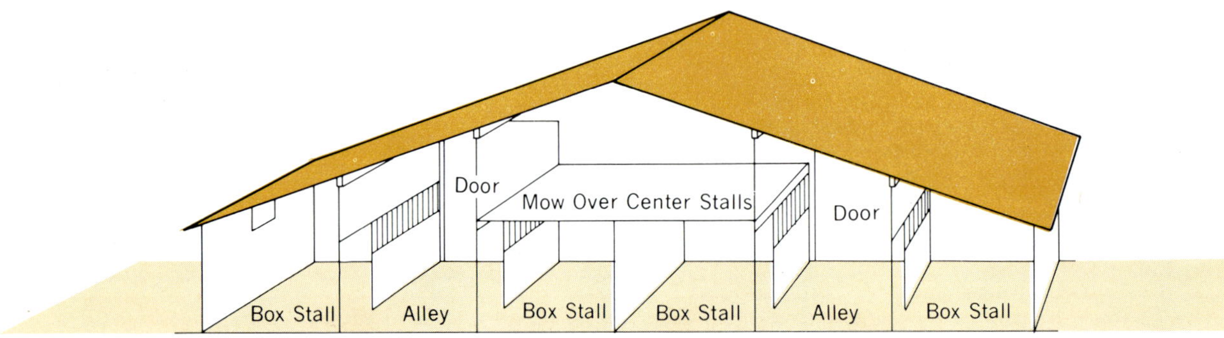

Four Rows of Box Stalls Serviced From Two Alleys. Two Center Rows Of Stalls Back To Back With Mow Above (post & beam)

Gable Roof

The gable roof shape is the most widely used for both open front and closed buildings and for other horse housing facilities. It is applicable to both narrow and wide barns.

The gable is a triangular roof with two equal pitches that meet at a line in the center of the span. It has a pleasing appearance, is medium in cost, and is fairly simple to construct and insulate. It is adaptable for natural overhead lighting through fixed translucent roof panels, and for natural building ventilation through eave and ridge openings.

Side extensions for roof overhang, covered way, or greater barn width can be obtained by increasing the roof height of the basic building without changing the roof pitch, or by changing the roof pitch within the limits of good design. Headroom depends on wall height, roof pitch, and type of framing.

The ceiling may follow the slope of the roof or be dropped to wall height, with or without overhead storage.

Framing

Common wood rafters supported on walls, or interior posts and beams, are widely used for buildings up to 72' wide. Rafter spans are influenced by stall size and service alley widths, and are most often 10'-16'.

Wood trusses are used extensively for clear spans to 60'. Pitches of less than 4/12 are seldom used.

Rigid frames, commercially produced of steel or laminated wood, are available for clear-span facilities to 100' wide.

Offset Gable Roof

The offset gable is a triangular roof with two equal slopes of different lengths meeting at a ridge. It resembles the combination roof, which is two unequal slopes of different lengths. The offset gable roof is used on both narrow and wide structures for horse housing and associated activities.

Framing

Common framing methods involve rafters supported on walls, interior posts and beams, or a clear-span truss or rigid frame gable roof with a roof extension on one side. The roof extension may be either clear span or supported on interior posts and beams. Both wood and metal construction are used.

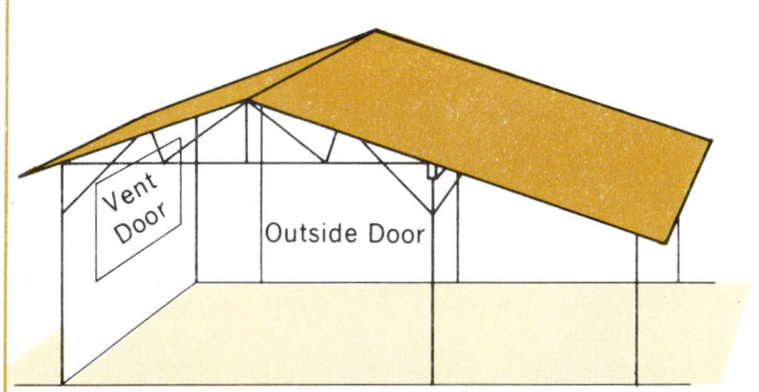

Open Front Shed (truss-post & beam)

Offset gable roof

Gambrel Roof

The gambrel roof has two slopes on either side of a center ridge. The lower slope is steeper than the upper slope. It is used mainly on two-story barns up to 40' wide.

Framing

The most common type of framing is light, clear-span braced rafters, supported on the barn wall and anchored to the mow floor joists to resist horizontal forces and uplift.

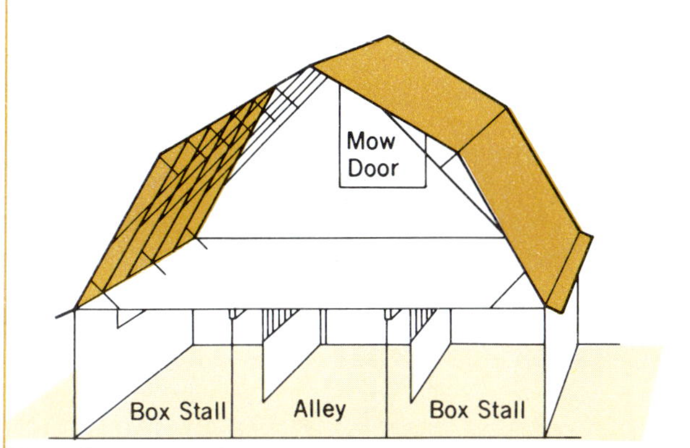

Two Rows Of Stalls Serviced From Center Alley (clear span, braced rafter)

Gambrel roof

Monitor Roof

The full monitor roof is made up of two shed sections and a gable section for the center portion of the roof. The vertical wall area separating the shed and gable roof provides natural light and ventilation. The monitor shape is seldom used for horse buildings less than 36' wide. The center section may be wider than the sheds or the reverse may be true. The roof of a large gable structure with attached sheds is, in a sense, full monitor. Its use matches that of clear-span buildings of similar size with side extensions.

Framing

See Framing under shed and gable roof sections.

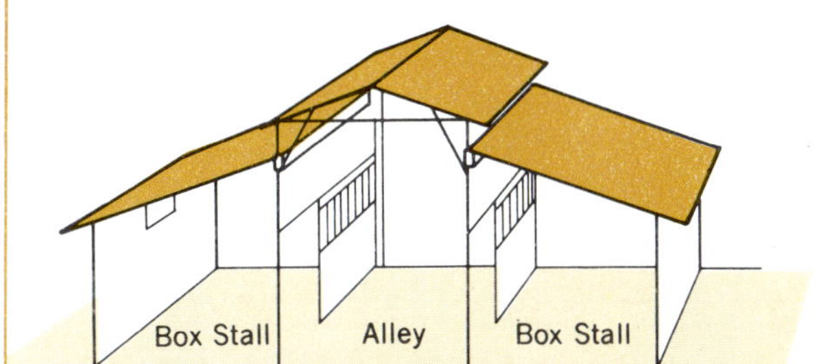

Two Rows Of Stalls Serviced From Center Alley (post & beam)

Monitor roof

Gothic Roof

The gothic roof is a pointed arch formed by two similar curved roof sections meeting at a center ridge. It is used on two-story barns and also for separate free-standing structures to provide shelter at ground level. It is adaptable to both narrow and wide structures.

Framing

Clear-span laminated or segmented rafters are supported by and anchored to the walls and mow floor, or to the foundation. Segmented wood rafters are usually for spans of 40' or less. Laminated wood rafters are suitable for almost all spans.

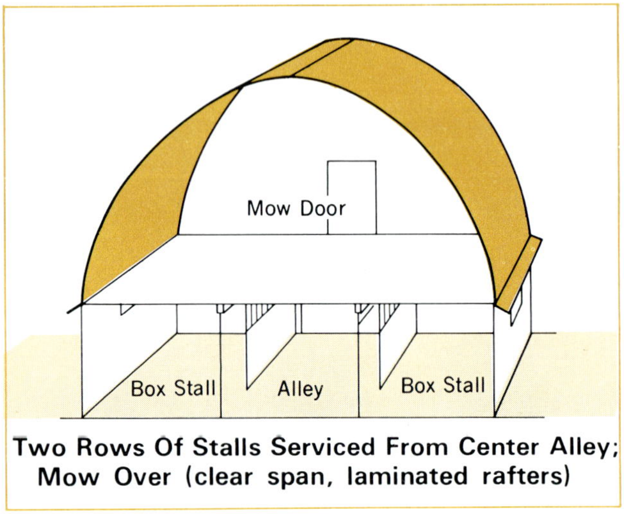

Gothic roof

SPACE REQUIREMENTS

Tables 1 and 2 list dimensions for basic facilities.

An open-front shelter is usually planned for 60 to 80 sq ft/1000 lb of animals housed.

Minimum ceiling height is 8' for a horse, 12' for horse and rider.

Doors are commonly:	width x height
Stalls	4' x 8'
Small wheeled equipment	10' x 10'
Horse and rider	12' x 12'
Large equipment, horse + rider	16' x 14'

Table 1. Space requirements for horses in buildings

	Dimensions of Stalls including Manger		
	Box Stall Size		Tie Stall Size
Mature Animal (Mare or Gelding)	10' x 10'	small	
	10' x 12'	medium	5' x 9'
	12' x 12'	large	5' x 12'
Brood Mare	12' x 12'	or larger	
Foal to 2-year old	10' x 10'	average	4½' x 9'
	12' x 12'	large	5' x 9'
Stallion[1]	14' x 14'	or larger	
Pony	9' x 9'	average	3' x 6'

[1]Work stallions daily or provide a 2-4 acre paddock for exercise.

Table 2. Hay manger and grain box dimensions (dimensioning inside the stall)

	Hay Manger[1]	Dimensions[2]	Grain Box
All Mature Animals (Mares, Geldings, Brood Mares, Stallions)	30" - 36"	Length	20" - 24"
	38" - 42"	Throat Height	38" - 42"
	20" - 24"	Width	12" - 16"
	24" - 30"	Depth	8" - 12"
Foals and 2-year olds	24" - 30"	Length	16" - 20"
	32" - 36"	Throat Height	32" - 36"
	16" - 20"	Width	10" - 16"
	20" - 24"	Depth	6" - 8"
Ponies	24"	Length	18"
	32"	Throat Height	32"
	18"	Width	10"
	20"	Depth	6" - 8"

[1]Wall corner hay racks are often used instead of mangers. Five feet is the usual distance between the floor and bottom of the rack. Many horsemen feed hay on the stall floor in both box and tie stalls and use a wall-mounted grain box in the corner of the stall.

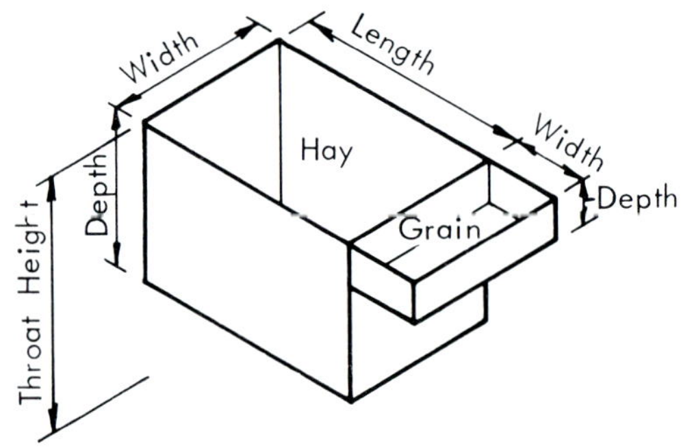

OPEN-FRONT BUILDINGS

When horses are kept outside, provide free-choice shelter. An open-front shed will usually provide adequate protection. If a large building is needed, provide a building at least 32' wide for moderate climates and 40' wide for cold climates, and large enough to provide 60 to 80 sq ft of floor space per 1000 lb of animal weight using the shelter. Additional room may be needed in the shelter for hay and bedding storage, foal creep, pens, etc.

Face the open side of the building, like the open side of any other livestock shelter, away from prevailing winds. The desirable clearance on the open side is 10'. Although hay and bedding storage may be in the shelter, feed hay outside the building in racks to conserve bedding. Provide an adequate supply of fresh clean water. This may require a heated stock waterer for winter use. A single heated water bowl will serve 8 to 10 horses.

When selecting a site for the shelter, consider its location in relation to headquarters buildings and traffic lanes for the convenience of handling animals and servicing the facilities. Some artificial light is desirable. Provide 100 watts of light per 500 sq ft of floor area. Install ridge devices, and adjustable panels in the back wall, for summer and winter ventilation.

BARN LAYOUTS

Barn layouts are associated with both function and engineering design. Barns classified by function or use are: breeding, riding, loose housing, hay, and utility.

Frame, pole, masonry, arch, rigid frame, etc., are classifications based on engineering design. To provide a more detailed description of barn types, classifications are frequently modified or expanded by reference to other design features associated with the building superstructure or use. Common reference features are: barn width and length; construction materials; wall construction; roof framing; ceiling height; wall and roof bracing; anchorage of superstructure to the foundation; floor materials; stall types and sizes; and service facilities for animal care.

Barn Width

Determine the proper barn width during the planning stages. Match the functional requirements of the building with the attainable wants of the owner to determine the width (and length) of a structure. The outside dimensions of the structure should allow for all of the planned inside arrangements and facilities without compromising on size.

Outside dimensions of buildings always indicate the outside size of the finished structure. Some space in most buildings is taken up by walls and other structural members, reducing the usable space inside the building. This reduction may result in less space and smaller facilities than those actually needed, especially if planning decisions were based on minimum space requirements, and the space-consuming structural features of the building were

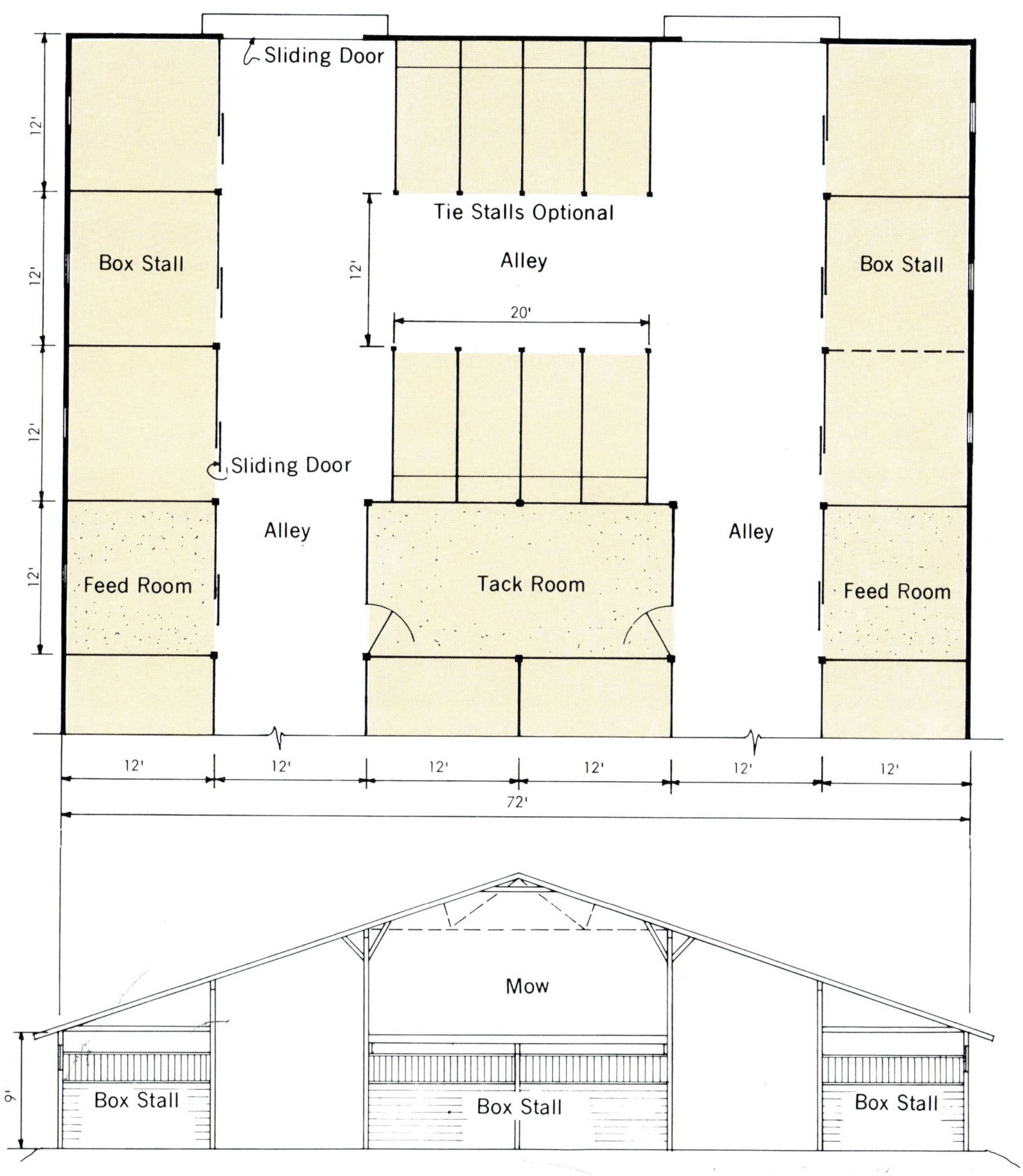

Four rows of stalls, two alleys (inside service)

ignored. Two rows of 10'x10' box stalls separated by a 10' wide alley will not fit into a building having a 30' outside dimension; but, they will fit nicely with very little room to spare in one that has a 32' outside dimension.

Check and recheck inside measurements against space requirements and facility sizes before selecting a building plan.

Minimum space requirements cannot be reduced when the welfare of the animal and the performance of associated activities are considered. However, reference to minimum dimensions does not preclude more expansive facilities. Many horse owners consider minimum dimensions and space requirements undersized and incompatible with adequate housing, horse care, riding enjoyment, convenience, and the efficient use of labor.

Layouts For Box Stalls

The width of a structure for box stalls is determined by stall size and alley width. Typical building widths (outside measurement) for enclosed structures and for various box stall arrangements are illustrated. The most often used width increment is 12'. Clear distances inside stalls and alleys will then be about 11'-11'-6". Select actual sizes to meet your needs.

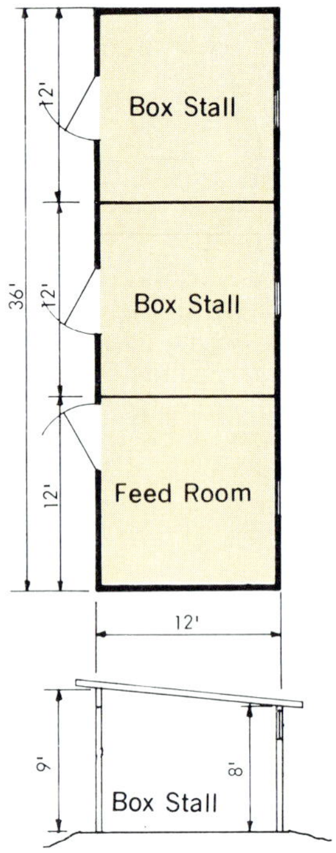

Basic facility for two horses (outside service)

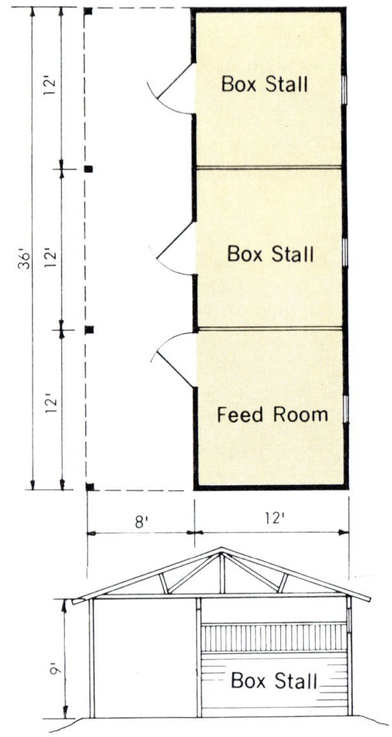

One row of stalls, covered way (outside service)

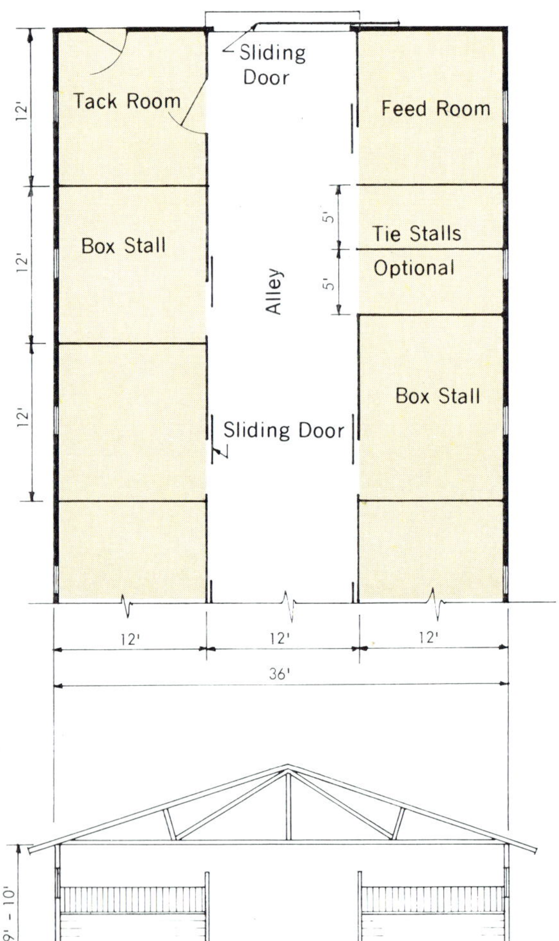

Two rows of stalls, center alley (inside service)

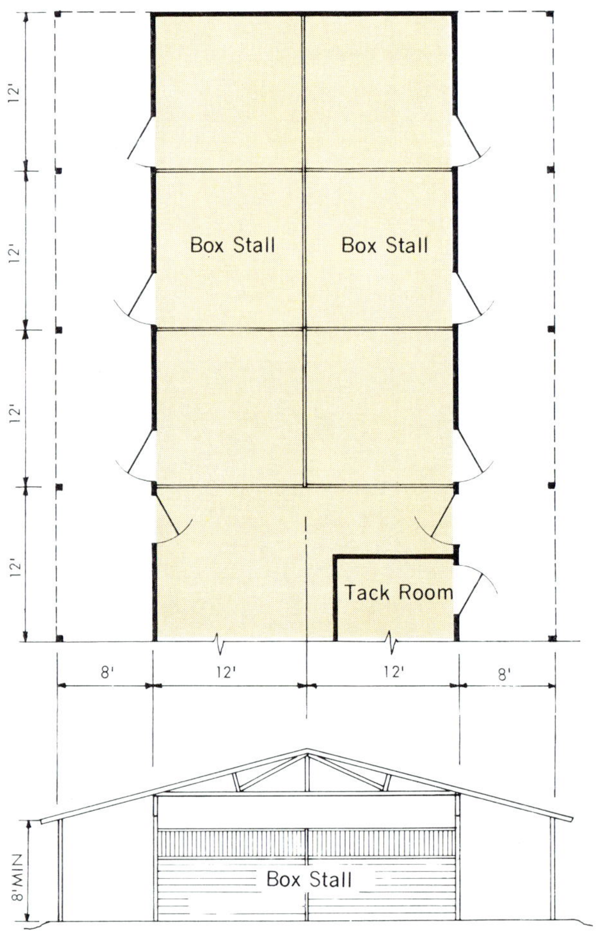

Two rows of box stalls (outside service)

Layout For Tie Stalls

The width of a structure for tie stalls is determined by stall length and alley width. If the building has both box stalls and tie stalls, the size of the box stalls usually determines barn width.

Typical building widths (outside measurements) for enclosed structures and various tie stall arrangements are illustrated.

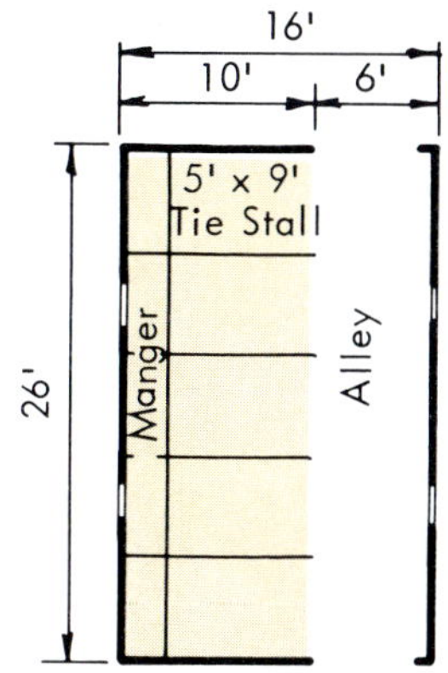

One Row, No Feed Alley

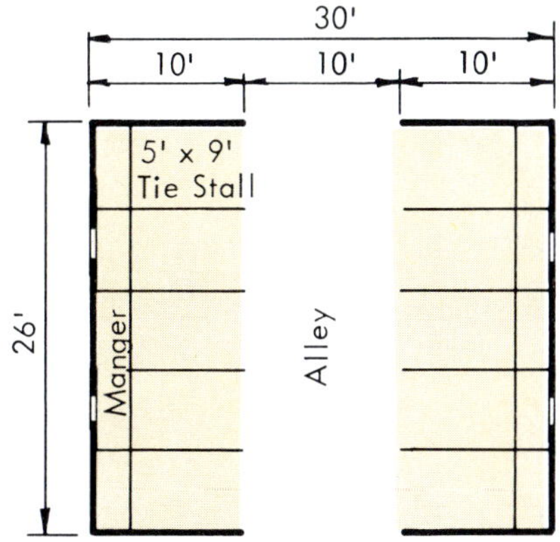

Two Face-Out Rows, No Feed Alleys

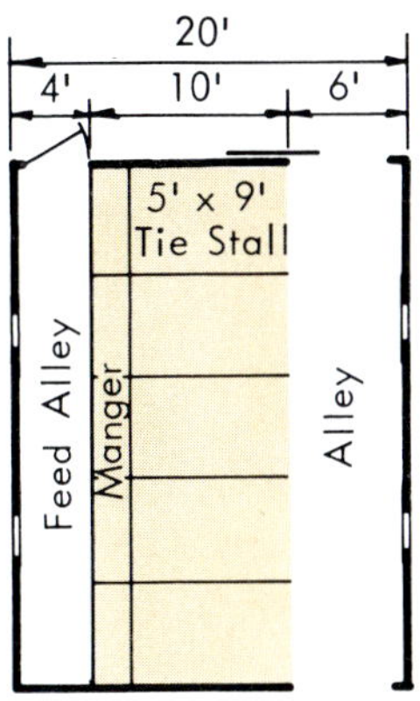

One Row, Feed Alley

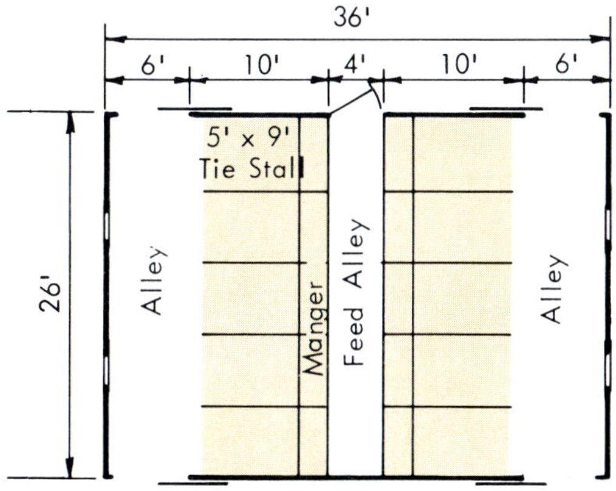

Two Face-In Rows, Feed Alley

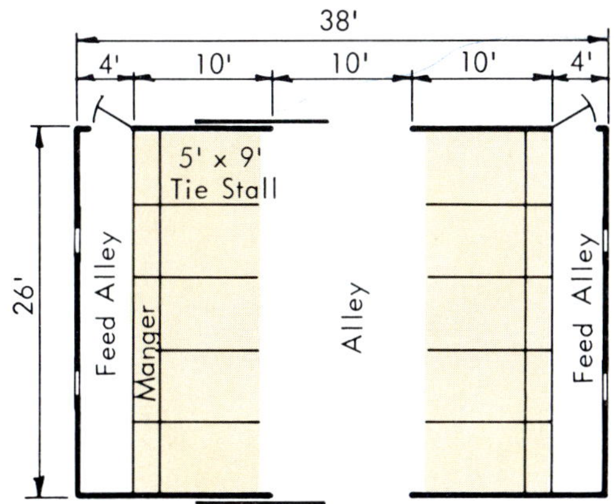

Two Face-Out Rows, Two Feed Alleys

Layouts For Other Horse Facilities Are Variable

The important thing is to have enough clear width (as well as ceiling height and building length) for the activities associated with the structure. Activities involving strenuous exercise are limited to wide clear-span structures or outside arenas. The various covered facilities and their widths based on use are:

- Alleys separating two rows of stalls in box stall barns are used for handling and resting animals, and for limited exercise and training. The alley may be as narrow as 10' in barns 32' wide, and as wide as 28' in barns 50' wide.
- Basic clear span structures 36'-50' wide are used for exercising and training horses and as sales barns. Riding is limited.
- Basic clear span structures 50'-100' wide are used as exercise, training, and riding arenas. Widths of less than 60', though usable for riding, are not as desirable as widths of 60' or more. Widths of 80' or more are considered spacious. A clear span width of 110' is recommended for an indoor show ring.
- Roofed, open-front housing structure widths are variable. A width of 32' is considered minimal for animal shelters, storage and utility buildings. Clear span structures are usually more desirable than post and beam construction.

The existing building site and prevailing management practices often influence both building width and length. Many basic wide-span buildings have attached sheds or roof extensions for stabling horses, storage and utility purposes, or bleacher seats.

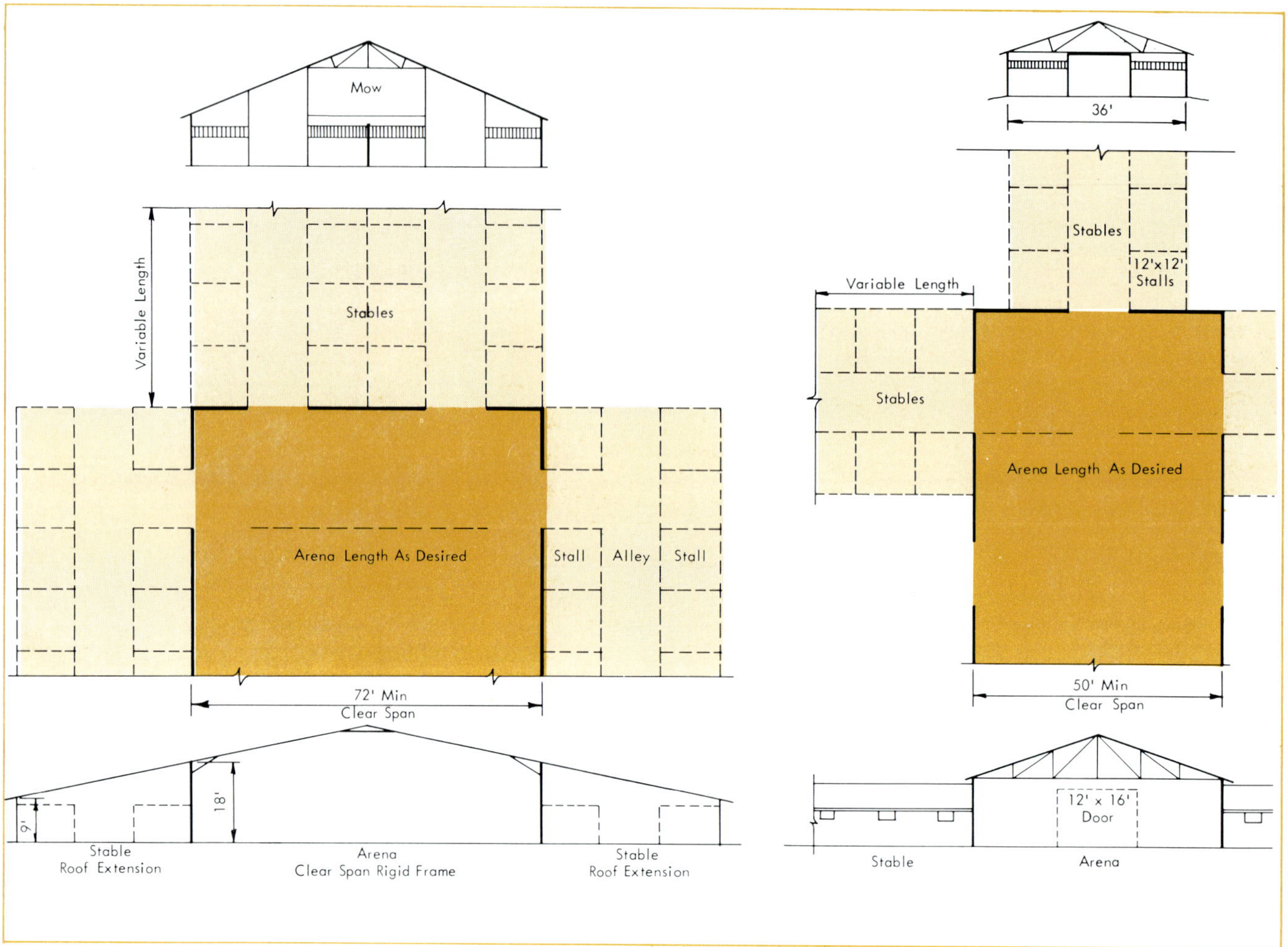

Ideas for an arena-stable combination with controlled access. Stables & arena may be separate buildings.

BARN CONSTRUCTION

Ceiling Height

The minimum distance from the floor to the under side of floor joists in a two-story barn, or to the under side of the lower chord of trusses or rafters in a one-story barn, should be 8' for the horse only, and 12' for the horse and rider. Low ceilings interfere with ventilation, make the barn dark, and are a safety hazard for man and horse.

Commonly used ceiling heights for stall barns are 9'-10' and for riding areas 14'-16'.

Alleys

For convenience, safety, and animal traffic, provide 10' or wider litter and work alleys between stalls. This allows room for moving horses, a small truck, or a tractor pulling a wagon or a manure spreader.

Provide at least a 6' litter alley when it is back of a single row of stalls. More hand labor is required to clean narrow alleys; also, moving horses through them is more difficult.

Use a 4' minimum width for feed alleys for convenient feeding, especially with baled hay. Cross alleys are also 4'. The paving of alleys is optional.

Doors at the ends of litter alleys are either sliding or overhead type and should be sized for the kind of traffic using them.

Stalls

Table 3. Characteristics of stalls

Item	Tie Stalls	Box Stalls
Water	out-of-stall	in-stall
Feed	in-stall	in-stall
Manure	less carrying	more carrying
Bedding	less required	more required
Exercise	out-of-stall	limited in-stall
Space	45 to 60 sq ft	100 to 320 sq ft
Floor	clay or plank	clay or plank
Partitions	strong & tight	strong & tight
Top Guard	manger end	on partitions

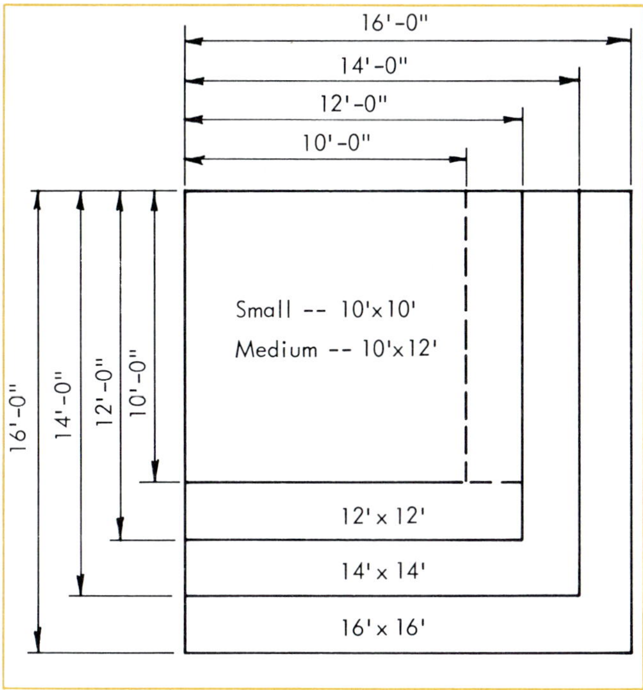

Box stall sizes

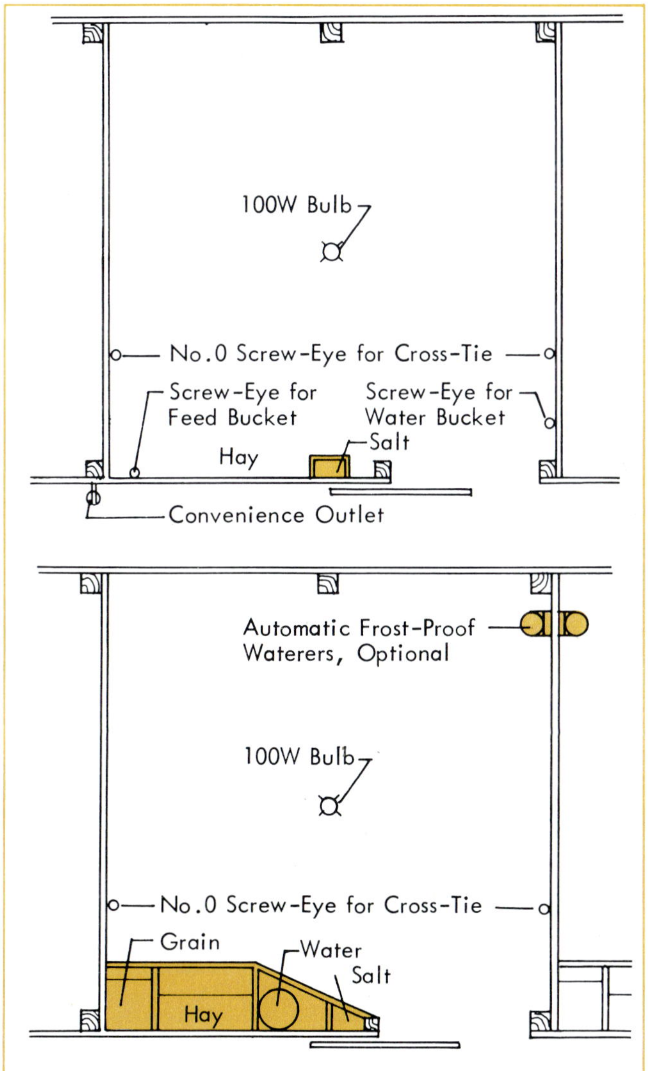

Typical box stall layouts

Many barns have only box stalls; but seldom does a barn have only tie stalls. In most horse barns some box stall space is necessary for sick animals, mares at foaling time, and colts.

Use stall walls that are tight, smooth, and free of sharp corners, loose wires, protruding bolts and nails, and anything else that might injure the horse as it moves about and lies down. The walls must be flush with the floor, so a horse cannot get its feet under the partition.

Wood is the most common material for stalls—commonly 2" hardwood planks to withstand rough treatment, chewing, and kicking. Concrete masonry walls are widely used in areas where climatic conditions can be controlled by waterproofing the outside stall walls and filling the block with insulation material to reduce heat loss. Block are smooth and withstand rough treatment, and chewing by the horses is no problem. Concrete masonry partitions in horse barns are becoming more common.

Box stalls

For riding horses, the minimum box stall is 10' x 10'. More commonly, box stalls are 12' x 12', although stalls 16' x 16' or larger are not uncommon. If the barn layout permits, a stall 16' x 20' or larger is useful for foaling mares. Box stalls for ponies may be smaller, depending on breed. A large stall can be obtained by removing the common partition between adjoining box stalls.

Equip box stalls with rugged Dutch doors, a full length sliding door, or swinging gates hung with heavy hardware. Horses can open doors, so to be safe, install door fasteners that the attendant can operate from inside and outside the stall and that the horse cannot open. A stay roller or guide is required outside the bottom of sliding doors to hold them in place.

Provide a convenient and safe stall arrangement. Consider both the attendant and the horse. Include provisions for watering; feeding hay, grain, salt and minerals; and for cross-tying the animals.

Either built-in hay-grain box combination mangers similar to those used in tie stalls, (page 16), or individual wall-mounted hay racks and grain boxes, are used in box stalls. Or, hay may be fed on the stall floor.

The stall corner supports are usually pressure-treated posts or poles; steel pipe up to 5" in diameter is also used. In post and beam construction, the posts or poles serve as anchors for the end of the stall partitions, and their size is based on roof and mow design. All wood in contact with the ground or foundation should be pressure-treated.

Box stalls are preferred to tie stalls for pleasure horses. Use 7' high box stall partitions to prevent fighting. The lower 4½'-5' of the 7' stall partitions and inside walls are usually made tight in cold climates, to prevent drafts and to protect the animals. Partition boards, spaced not more than 1½" apart, allow better ventilation but may require vertical center bracing to keep the wall true. Concrete masonry partitions are popular because of ease of construction, durability, and low maintenance.

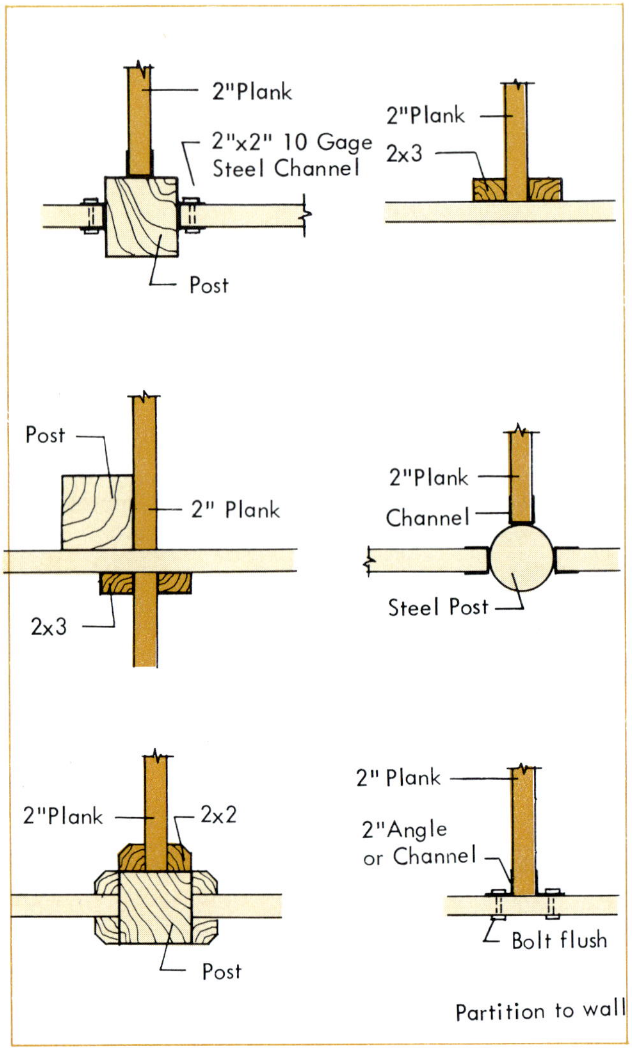

End anchors for stall partitions & walls

The upper portion (2'-2$\frac{1}{2}$') of the stall partition is often a rugged open panel or guard. This more-or-less open section aids ventilation and makes observation possible. The alley wall may be 5$\frac{1}{2}$' high with no guard.

Partition heights

Various materials are used for the open guard, such as: vertical $\frac{1}{2}$" diameter steel rods, or $\frac{3}{4}$" pipe spaced not more than 4" on center; sections of #4 gage welded steel fence; flattened 1$\frac{1}{2}$", 9 gage expanded steel mesh; #6 gage welded wire 4" x 4" mesh; chain link fencing; vertical wood slats.

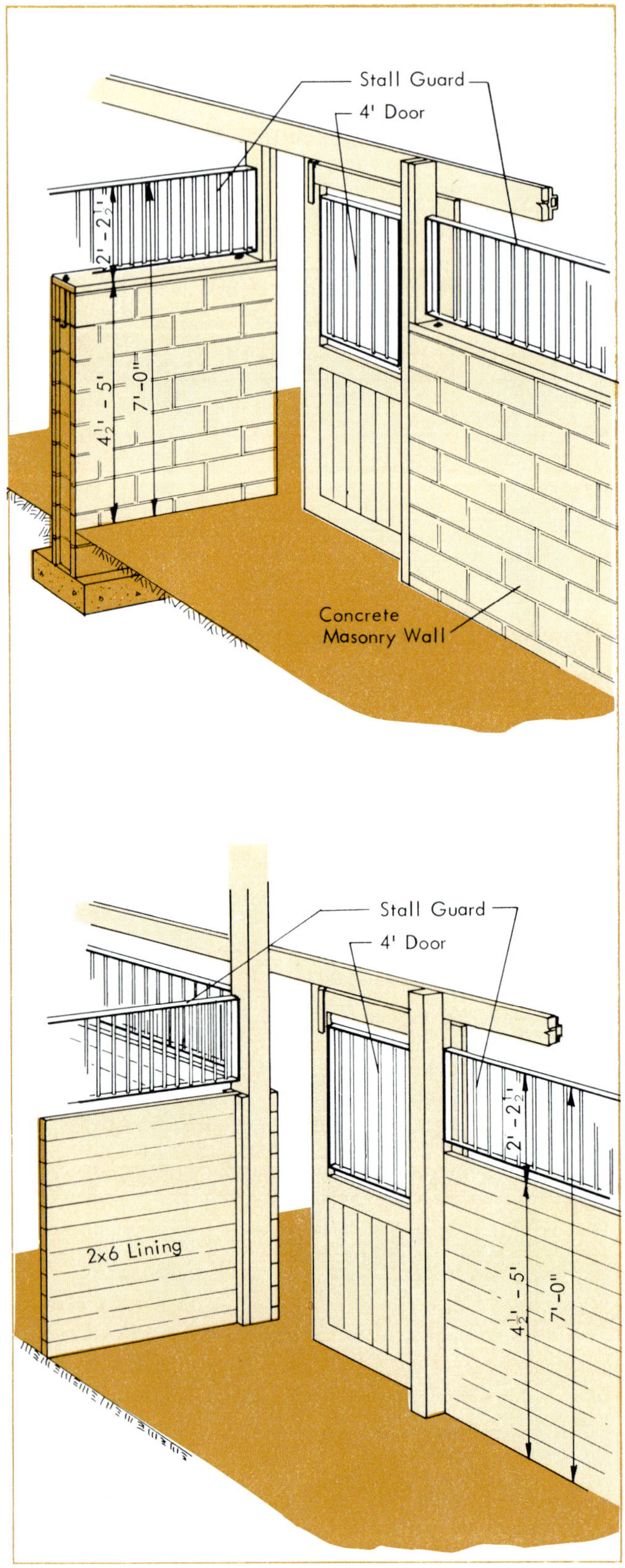

Typical box stall construction and arrangements

Tie Stalls

Tie stalls are not as widely used for pleasure horses as box stalls. However, they do provide an area for confining animals that are restless in box stalls. Tie stalls require about half the area, less bedding, and are easier to clean than box stalls. They can often be constructed in existing buildings unsuitable for box stalls.

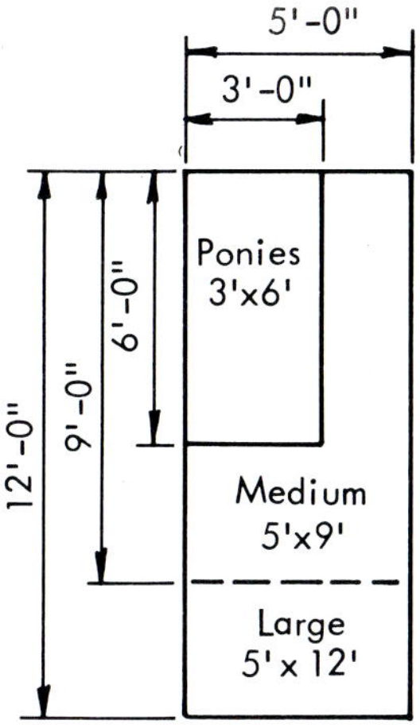

The typical tie stall is 5' x 9' (3' x 6' for ponies), although stall lengths up to 12' are often used. The length of the stall is measured from the front of the manger or grain box to the rear of the stall partition. The top of a wood built-in manger is about 2' wide. The standing platform is 7' long or longer. The built-in manger slopes forward at the bottom on the standing platform side to protect the stabled animals' knees. Individual grain boxes and hay racks are sometimes used in tie stalls instead of built-in wood mangers. Hay is occasionally fed on the floor. Horses are usually removed from the stalls for water and exercise, although watering in the stall is not uncommon.

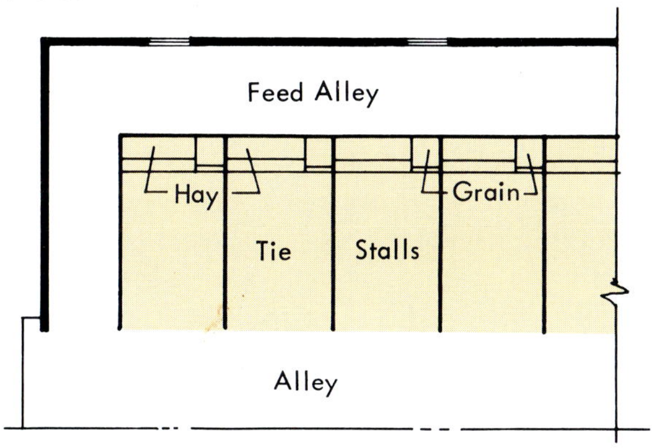

Tie Stalls facing out - serviced from front feed alley - Built-In Manger

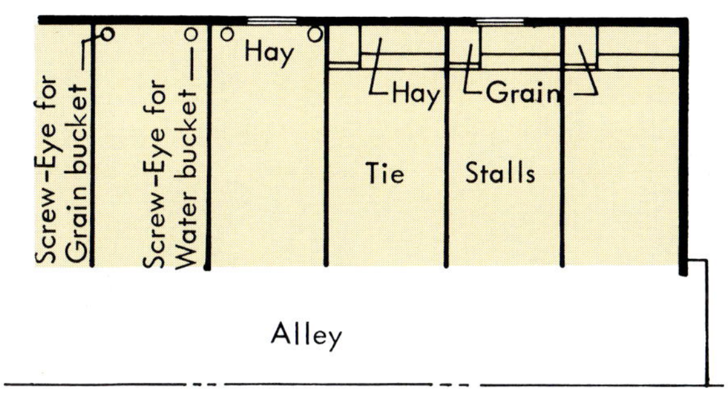

Tie Stalls along wall - serviced from the rear Built-In Mangers & Minimum stall equipment shown.

Typical tie-stall arrangements

A litter alley at least 6' wide and with a gutter is required behind a single row of tie stalls. A 10'-12' litter alley with gutters separates two rows of stalls facing out. A 3½'-4' feed alley in front of each row of stalls is desirable. But, feeding can be handled satisfactorily from the rear of a stall not serviced by a front feed alley.

Provide partitions between tie stalls high enough at the manger end to keep the animals from fighting. Use a height of at least 6' for the front or manger end of the partition, and 4½' at the rear.

The amount of bedding needed depends upon the type of floor, type of stable, and the weather. Provide 8 to 15 lbs of bedding per day per 1000 lb animal weight in occupied stalls. Always keep the bedding clean and well fluffed up. Never leave it hard or lumpy.

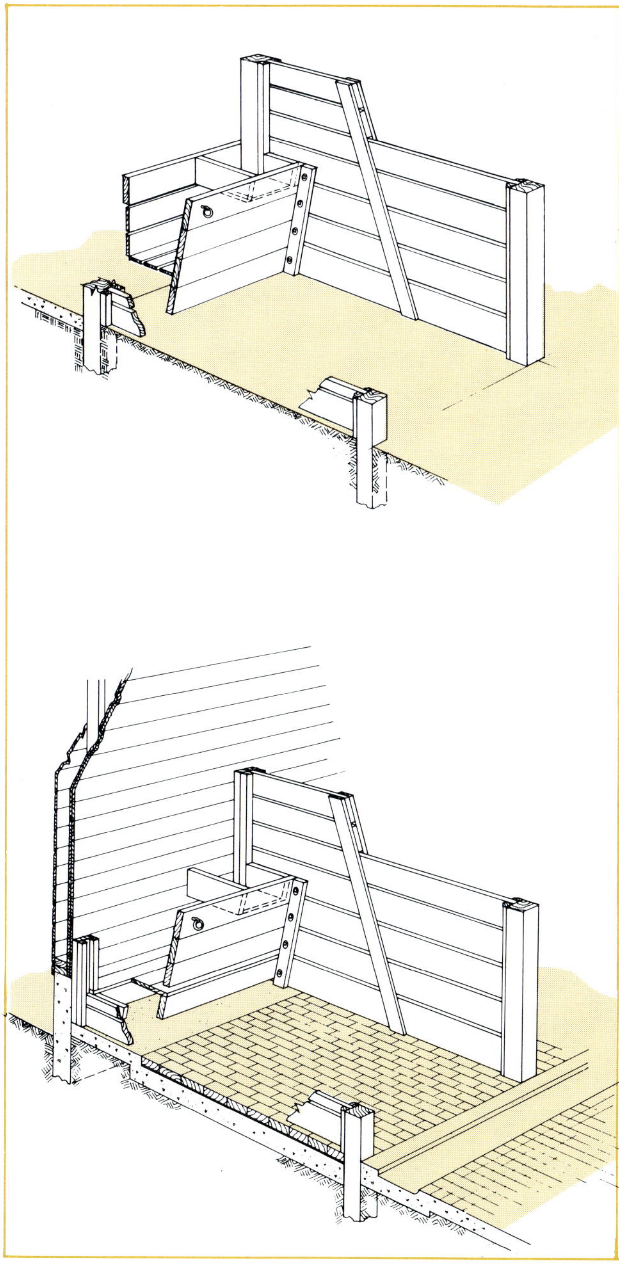

Tie stalls

Tie-stall details

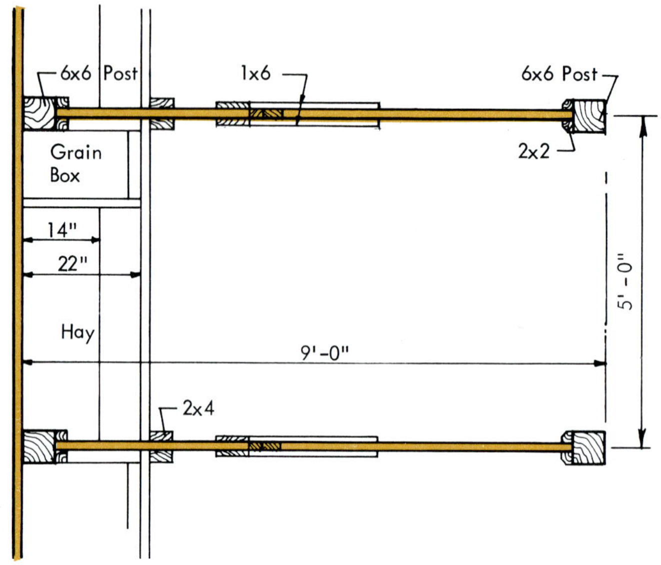

Floor Plan

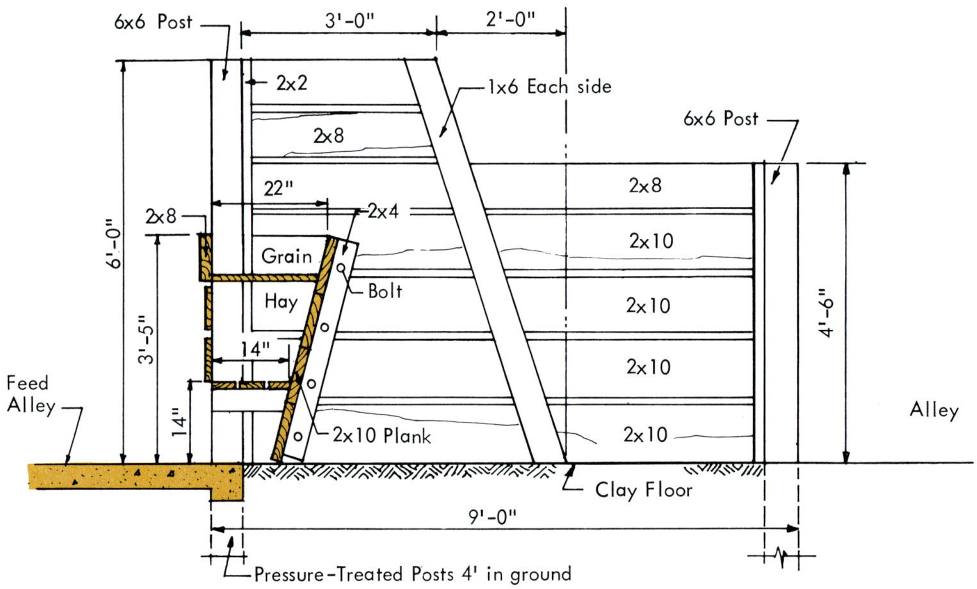

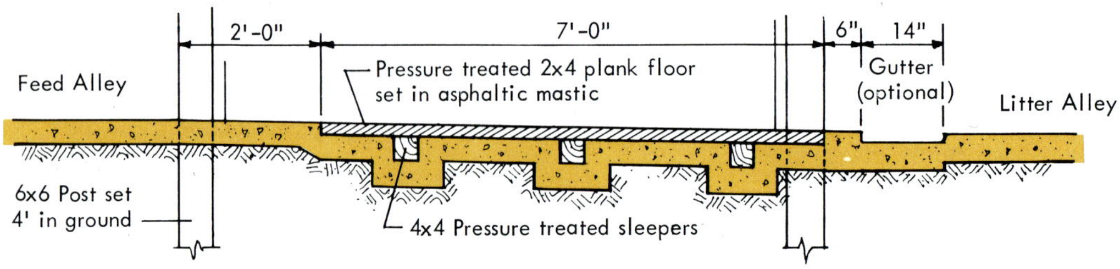

Alternate Floor

Feed Room

Organize the feed room, which is seldom larger than the size of a box stall, for convenience and easy housekeeping. Plan storage for feed materials, equipment, and tools. Provide an uncluttered traffic pattern to reach the stored materials without interference. Keep the storage area as dust free as possible.

Hay may be stored in an overhead loft or on the ground floor. Small rooms and narrow doors are inconvenient for storing hay and add to the chore labor required to handle it, both in and out of storage.

Grain feed must be stored in vermin-proof bins or containers. Metal containers such as large garbage cans and wood bins with tight lids are excellent for limited bulk feed storage. Hopper-bottom bins with mechanical unloading are also used either inside or outside the barn. Homemade and commercial grain bins are used for storing grain in large quantities.

The feed room should be well lighted, and equipped with convenience outlets. Provide 150 watts of light and at least two convenience outlets in the feed room. The entrance door to the feed room should be at least 4' wide and equipped with a latch so that a loose horse cannot get into the feed. A good rodent control program is essential.

Provide space for feed storage on a rule-of-thumb basis: 1¾ lb of hay and 1 lb of grain per day per 100 lb of the horses' weight. Baled hay will weigh approximately 10 lb/cu ft, and a ton will occupy approximately 200 cu ft. Whole oats weigh approximately 26 lb/cu ft (80 cu ft/ton); ear corn about 28 lb/cu ft (70 cu ft/ton); and bulk feed about 40 lb/cu ft (50 cu ft/ton).

The amount of storage space needed will depend on the number of animals, their total weight, and how often the feed supply is to be replenished.

Tack Room

A well organized and maintained tack room that is enclosed, dry, and free of dust is important to good stable management. In smaller stables it is simply an equipment storage area. In larger stables the tack room has traditionally been the horseman's headquarters for essential equipment and activities associated with horse owners. It can be simply a small area or room for riding equipment, or it can be large enough to serve as an office, a service shop for cleaning and maintaining tack, and a meeting and lounging place for the clan.

Build the tack room large enough for the activities associated with it. The tack room may contain all or some of the following equipment and facilities: saddle racks, tack box, bridle and halter racks, shoeing box, first aid kit, clothes closet, storage cabinets, shelves, filing cabinets, desk, chairs, miscellaneous furniture, refrigerators, sink, working and loafing areas, heating equipment, hot and cold running water, and cooking facilities. Toilet and shower facilities may be adjacent.

Install at least 300 watts of incandescent lighting (100 watts fluorescent) per 100 sq ft of floor area and provide adequate convenience outlets.

The tack room provides a warm area in the barn when insulated and heated for winter use. Side wall and ceiling insulation 2" thick plus electric heat and ventilation economically keep the area dry and comfortable in cold weather.

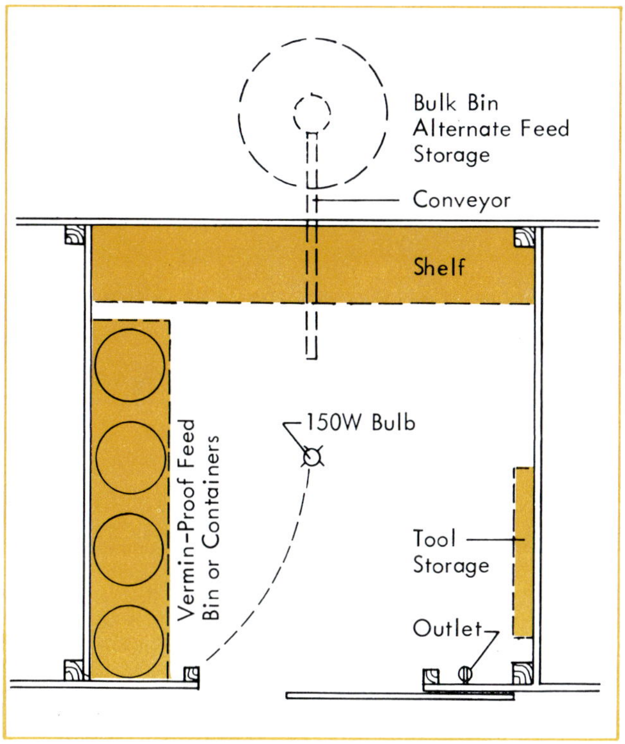

Typical feed room

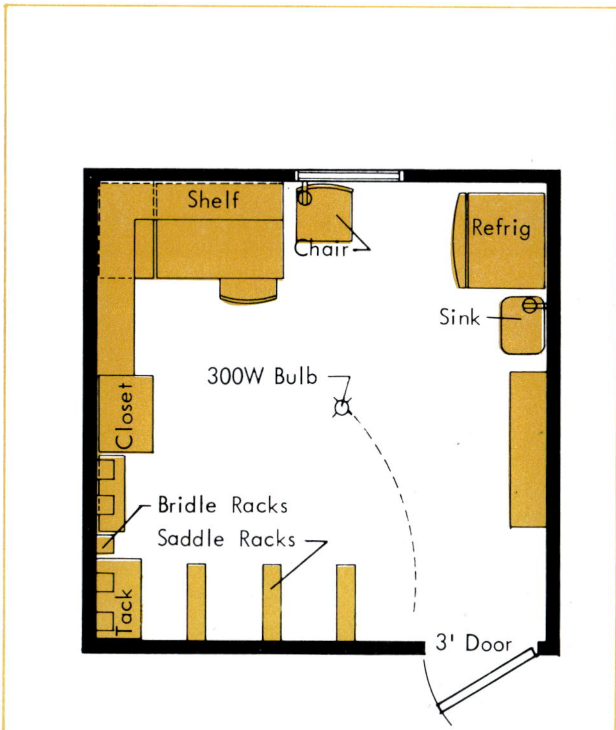

Typical tack room

Floors

Packed or puddled rock-free clay on a well-drained base makes one of the best floors for stables. It is usually easy to obtain. However, it is difficult to keep clean and has to be renewed from time to time. Wood plank stall floors or wood block floors on concrete are preferred by some horse owners, but such floors are very difficult to keep dry and free of odors. Concrete floors are the least desirable, and if used, a considerable amount of bedding, such as sawdust, is needed to prevent stiffness. Artificial turf (tartan) is used for stall floors and alleys by some innovators.

To ensure adequate drainage, elevate the top of the floor 8"-12" above the outside ground level. Concrete floors are used in the wash area and are desirable in the feed room, feed alleys, and tack room. Avoid floor finishes that are too slick, such as troweled concrete and some asphalts. Examples of floor construction are shown in the tie stall construction details.

Mangers and Racks for Hay and Grain

Construct mangers and grain boxes from 2" hardwood or equivalent material using the dimensions in table 2, page 8 . Corner or wall hay racks may be used instead of built-in wood mangers. Corner hay racks extend approximately 2' each way from the corner of the stall and usually have a circular front. Wall hay racks are approximately 4' wide with straight or circular fronts. All mangers should be constructed so they are convenient for horses to eat from and convenient for attendants to service.

Separate grain boxes, usually located in a stall corner for convenient feeding, are used in both box and tie stalls, especially when hay is fed in wall racks or on the stable floor. Grain boxes in box stalls are usually filled through a small opening in the front wall, and in tie stalls from the rear of the stall or from a front feed alley.

Windows

A small adjustable 2' x 2' window in each box stall will provide light and assist in ventilation. However, windows are not essential if other methods are used for lighting and ventilating the building. Locate windows near the top of the wall. Allow at least 6' between the window sill and the stall floor. Protect all windows that can be reached by horses with heavy 1" x 2" welded wire or a steel grating. Attach the protective guard to the window so it can be easily removed for window cleaning.

Translucent roof and wall panels are used to admit daylight into some animal shelters. However, they may become a heat source in the summer and are subject to severe moisture conditions and frosting in cold weather.

Windows add to building costs. Summer ventilation can be provided by solid adjustable ventilation doors near the top of outside walls instead of, or in addition to, windows. Jalousie-type windows are widely used in horse barns—they admit air and light, exclude rain, are easily adjusted, and window guards do not interfere with their operation.

Lighting

Good illumination is important for the convenience and safety of both the horse and the attendant. Provide lighting for:

- General illumination of alleys and pathways.
- Specific illumination of stalls, storage areas, and specialty areas.
- Outside approach and service areas.

For stall lighting, place one 100 watt bulb near the center of each stall if height permits (9' or higher); locate switches near the stalls. Foaling stalls need about 300 watts of light.

Recess lights into the ceiling or use guards so the animals cannot break the bulb and fixture. Dust-free fixtures are desirable. Reflectors will improve lighting in stalls, alleys, and walk areas. Locate convenience outlets for the use of portable lights and equipment in alleys, tack room, and storage areas.

Locate general illumination at exterior doors.

Control outside lights from a switch near the barn entrance, or 3- or 4-way switches to allow control from both the barn and the residence.

Electric wiring and fixtures must comply with applicable state and national electric codes.

Water

A healthy horse needs an adequate supply of fresh water of the same quality as is used for human consumption. A mature horse needs 8 to 12 gallons per day. It is difficult to supply this much water by carrying it in pails, especially if several horses are to be watered.

The vast majority of horses are hand watered. Locate in-stall watering devices where spillage can drain directly from the stall—they may need filling more than once a day in a cold barn because of freezing. A stop and drain (frostproof) hydrant or an electrically heated watering bowl will supply fresh water in freezing weather. One electrically heated watering unit can serve two stalls or serve 8 to 10 horses in a lot. If horses are confined in tie stalls and a number are released at once, a small frost-free watering tank is desirable.

Bedding

Provide dry bedding that will adequately support the horse and keep it comfortable. Useful materials include peat moss, straw, shavings, sawdust, and old hay mixed with wood shavings.

Table 4. Bedding material approximate water-absorbing capacity

lbs of water/lb of bedding	
4.0	Tanning bark
3.0	Pine chips
2.5	sawdust
2.0	shavings
1.0	needles
1.5	Hardwood chips, shavings or sawdust
2.5	Shredded stover
2.1	Ground cobs
2.8	Oats, threshed
2.5	combined
2.4	chopped
3.0	HAY, chopped mature

MANURE HANDLING

Manure management is an important part of caring for horses. The physical and social aspects of collecting, handling, storing, transporting, and disposing of manure are a major problem. Problems occur when collection, storage, and disposal methods contaminate a water supply and are responsible for odors, flies, and other nuisances.

Check local regulations for storage and disposal recommendations. If regulations exist, follow them. In the absence of regulations:

- Dispose of manure daily when possible.
- Provide temporary storage for manure that cannot be disposed of daily—about 12 sq ft of fenced or enclosed storage per horse.
- Locate the storage in an approved or safe area for convenient removal, away from any water source and out of natural drainage channels.
- Empty the storage at least weekly during fly breeding season (spring temperatures above 65°F until the first killing frost in the fall).
- Drain the storage area away from surrounding facilities and lots.
- Keep all runoff that is polluted with animal waste from reaching usable or public waters by diverting it to a non-critical area or pond (lagoon).

FIRE SAFETY

Fire *prevention* is the best *protection*. Prevention includes adequate wiring, good housekeeping, properly maintained heating equipment, lightning protection, proper storage of fuels, and avoiding spontaneous combustion.

What To Do In Case Of Fire

- Make sure all persons are well away from the danger area.
- Call the local fire department. (Phone number should be posted near phone.)
- Rescue livestock, if possible, without risking human life.
- Close windows and doors to keep air supply to a minimum, especially if it is a major fire beyond the capacity of an extinguisher.
- Use the correct extinguisher to fight the fire.
- When the fire is out, be sure to ventilate the area to get rid of dangerous gases.

How Fires Are Extinguished

Three components are necessary to sustain a fire: heat, air, and fuel. Removing any one of these components will extinguish the fire. Fuels include those solids and liquids that will burn, and vapors from flammable liquids with low flash points. Since it is usually impossible to remove the fuel once the fire has started, fires are controlled by cooling the burning material, by cutting off the air supply, or both.

Provide Fire Protection

Portable fire extinguishers may control a small fire, or delay the development of a large one until help arrives. Every barn should be equipped with one or more portable extinguishers located near entrances and high-hazard areas. The need is greater in rural than in urban areas,because it takes longer for fire equipment to reach the location after the fire has been reported. Provide adequate water for fire department use from a farm pond located within pumping distance, or from a 10,000-gallon emergency cistern. Consult your local fire department officials for details.

Not all extinguishers can be used on any type of fire since various flammable materials respond differently to treatment.

Table 5. Fire classifications and recommended extinguishers

Class of Fire	Material	Extinguisher Recommended
A	Wood, paper, textiles, grass, trash, similar materials.	Pressurized water (dry chemical and carbon dioxide give temporary control.)
B	Grease, gasoline, oils, paints, kerosene, solvents	Multipurpose dry chemical, carbon dioxide, dry chemical foam.
C	Electrical equipment	Multipurpose dry chemical, carbon dioxide, dry chemical.

Consider the following when selecting an extinguisher:

- The size and type of fires most likely to occur. Some extinguishers can be used for more than one class of fire.
- Ease of handling. The operator must stand at a safe distance from the fire while operating it.
- Maintenance expense, frequency of inspection, and whether the extinguisher will be subjected to freezing temperatures.
- Only extinguishers which have been tested, rated, and approved by an accredited organization. Beware of cheap, simple devices which are claimed to solve all fire problems. Carbon tetrachloride extinguishers should not be used either inside or outside a building because they generate highly toxic gases when exposed to heat.

Water As A Fire Extinguisher

When a fire is discovered, the first impulse is put water on it. Be sure water is the material to use before using it. Water will put out some types of fires, but when used on others, the results may be disastrous, fatal, or both. For example:

Water will drown-out a Class A fire (wood).

Water may spread a Class B fire (grease), a type that has to be smothered out. Wet blanketing material may be used to smother a small Class B fire if it can be approached safely.

Water used on a Class C fire (electrical) may conduct current to the person fighting the fire, causing severe shock or electrocution. The end results are the same whether the fire is being squirted or smothered with wet blanketing material, whether the source of the water is a hose, fire extinguisher, or bucket brigade.

RODENT CONTROL

Kill Present Population

- Provide 15-25 bait stations:
  Two 1 x 10 x 6' boards to make a feeder for along side a building or fence row, or an inverted and weighted bushel basket.
- Locate bait stations at all buildings, lumber piles, and other rodent harbors.
- Keep stations stocked with clean, dry, poison bait. Mix dry anticoagulant concentrated poison with oatmeal or corn meal for bait.
- Follow directions on the container when mixing bait.
- After the first night, double the amount of poison in any station that was emptied. Continue doubling the amount offered until some bait is left uneaten at every station, then keep stations stocked with clean, dry poison. A rat, feeding on poison bait regularly, will die in 3 to 5 days.
- Fumigation will kill rodents in livestock buildings.

Clean Up Harbors

Force rodents to the bait station by removing outside shelter. After killing the present population, remove all hiding places—clean up junk, trim weeds from fence rows and around buildings.

Rodent-Proof Grain Storages

Force rodents to eat poison bait by preventing entrance to cribs, bins, and feed centers.

An 8" wide 28-gage metal strip, 3' off the ground and continuous around the building, will prevent rats from climbing the walls. Hardware cloth (¼" x ¼" mesh) from the top of the foundation to the metal strip will prevent rats and mice from eating through the walls. Flash (28-gage metal) around doors and other openings.

Rodent-proofing will not keep a grain storage free of rats and mice without a poison feeding program.

Maintain All-Year Program

One permanent feeder per building, checked every week, and kept stocked throughout the year with clean, dry, poisoned bait, along with a continuous clean-up program, will guarantee that migrants will be killed as they arrive on the farm.

WINDBREAK PLANNING

Orientation

Local experience is the best indicator on the distance facilities should be placed from shelterbelts.

Generally, shelterbelts should be 100'-300' away from protected areas. The shorter distance is suitable where snow accumulation is less severe.

Wind Patterns

With a 40 m.p.h. wind from the left, velocities will be reduced to about those shown. For other speeds the reductions will be proportional.

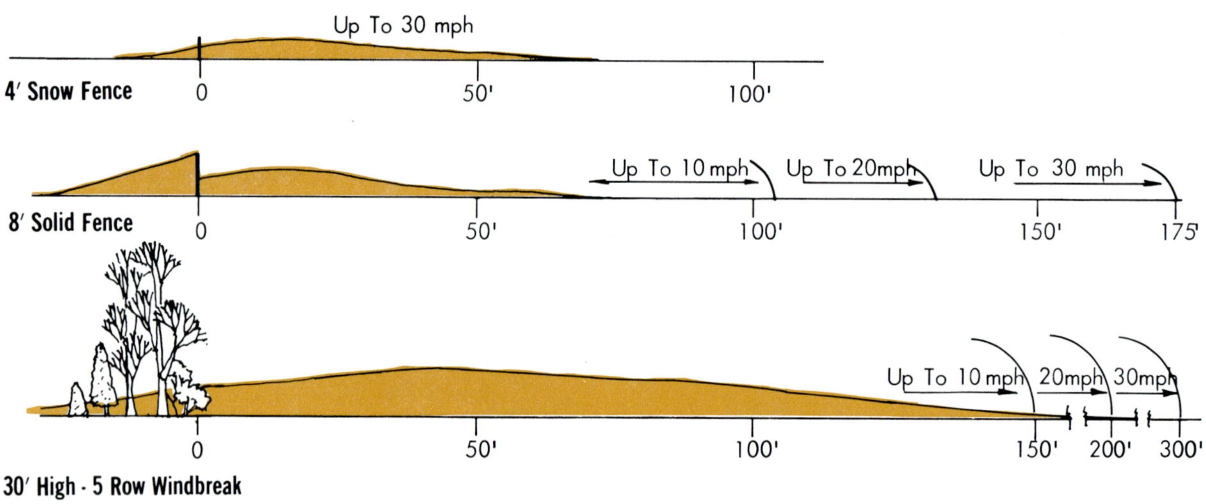

Typical Snow and Wind Patterns

Protecting Open-Front Buildings

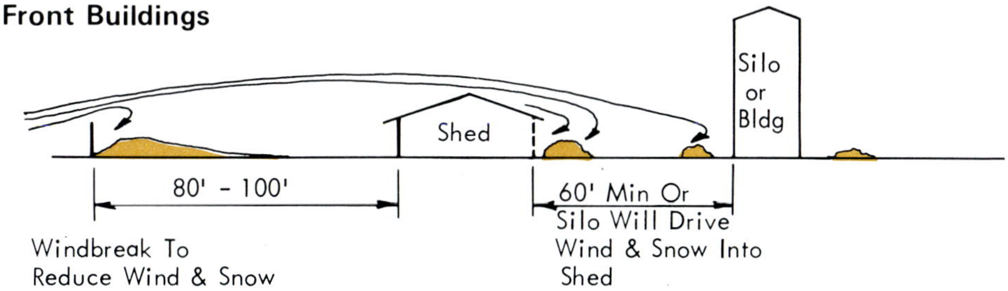

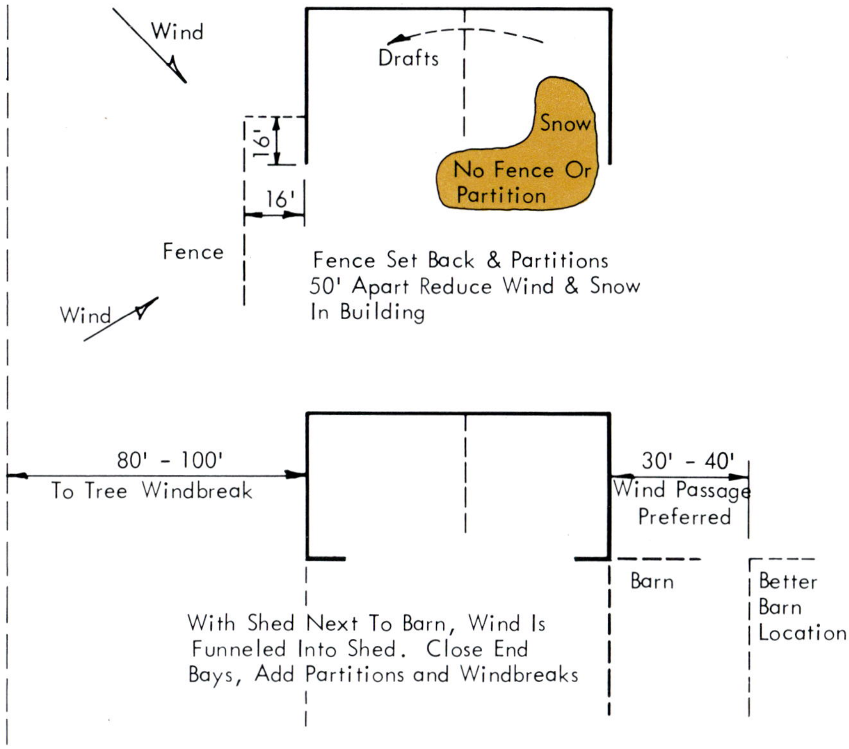

Solid Fence — Better Wind Protection

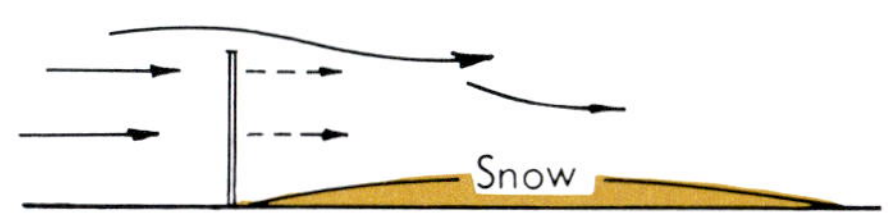

80% Solid Fence — Better Snow Protection

VENTILATION

Moisture is one of the biggest problems in all types of livestock housing. It can be controlled in a properly constructed facility. Horses, like other animals, produce heat and moisture. The moisture must be removed from the stable to prevent condensation and a build-up of odors. Ventilation is required to remove the moisture and odors, and to control temperature.

Install the ventilation system for the stable when the building is being constructed. This can be done by providing, in addition to doors:

- Eave openings, ridge ventilators, and adjustable windows or wall panels in both uninsulated and insulated buildings used for cold housing.
- Ventilating fans, fresh air inlets, adjustable windows or wall panels, and supplemental heat in insulated buildings used for warm housing.

NOTE: "cold" and "warm" housing refer to winter conditions inside the building. Temperatures inside *cold housing* will be nearly as cold as outdoors. In *warm housing*, temperatures are maintained at least above freezing.

As a rule, horse owners in mild or moderate climates can obtain adequate housing for their animals with uninsulated buildings. However, wood or masonry sidewalls and tight sheathing under the roof are desirable.

In northern regions, insulated buildings and supplemental heat are more commonly used to protect the animals and attendants from severe winter weather. Heated barns are also used for show horses to keep them in show condition throughout the year.

Enclosed buildings are usually tightly constructed with snug-fitting doors and windows. They may be uninsulated or insulated. Uninsulated and some insulated barns are used for cold housing. Only well-insulated buildings can be used for warm housing. As the move is made from cold to warm housing, the ventilation system requires more refinements.

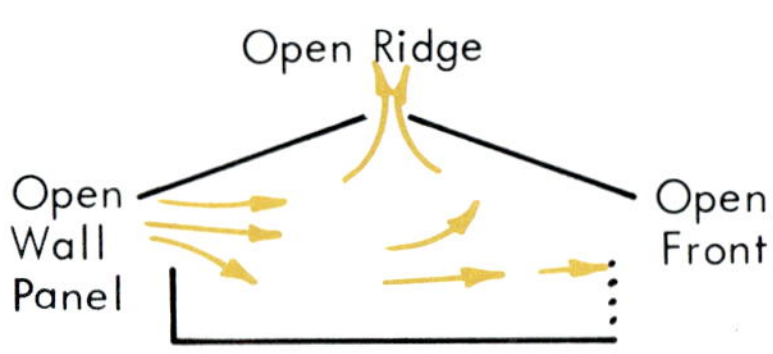

Summer

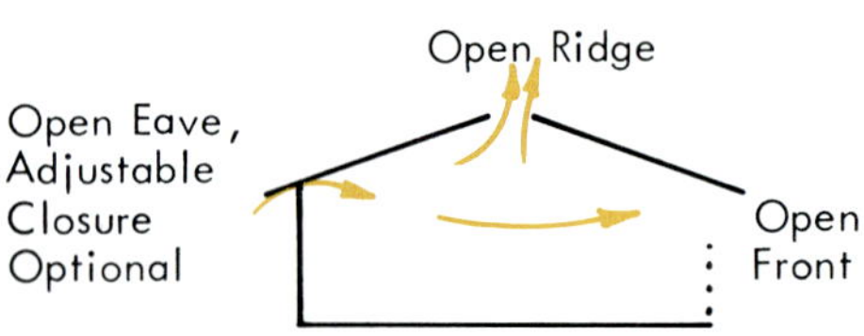

Winter

OPEN-FRONT BUILDINGS

The covered walls and roof of open-front buildings are usually of more or less tight construction. Sometimes insulation is used under the roof and on covered walls to reduce moisture condensation during the winter, and to modify temperatures in the summer.

An open-front building is designed to use natural air movement for ventilation; the temperature inside the building is usually a few degrees warmer in cold weather, and a few degrees cooler in hot weather, than the temperature outside the structure.

Open-front buildings do not require the ventilation refinements of enclosed buildings. However, the shelter must be provided with sufficient air inlets and outlets to permit the desired air movement throughout the building. Good air movement is needed in cold weather to prevent moisture from condensing on the underside of the roof, and in hot weather to keep the inside temperature acceptable.

Suggestions For Ventilating Open-Front Buildings

Locate the building on a well-drained site with the open side away from prevailing winter winds.

If drafts and snow blow-in are problems, close the front wall at each end and adjacent to solid partitions. Use spaced board windbreaks at the ends of the building.

Install solid cross partitions from floor to roof in long open sheds if indoor drafts continue to be a problem.

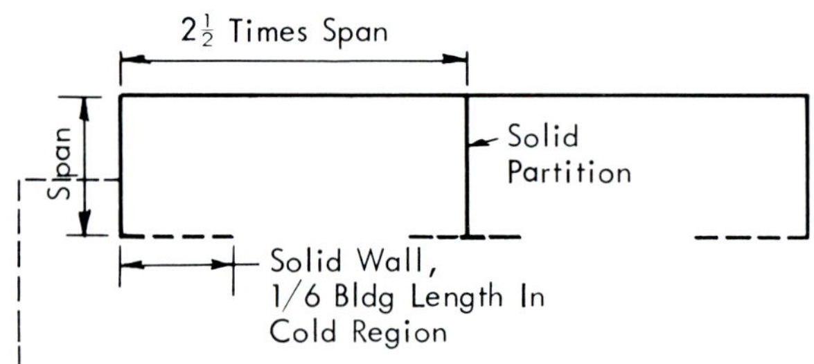

Spaced Board Windbreak Fence

Air Inlets

Provide small openings at the top of the back wall for cold weather air inlets. Add adjustable doors on the openings to prevent snow blow-in.

Provide large openings—doors or large panels—in the back wall for warm and hot weather inlets. About half of the wall should be openable.

Air inlets and outlets

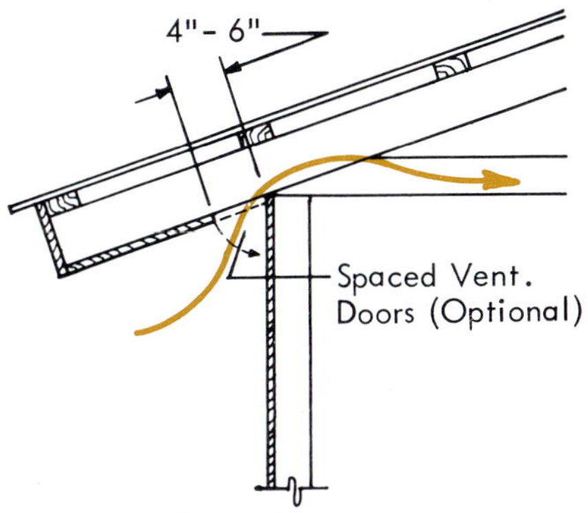

Cold Climate

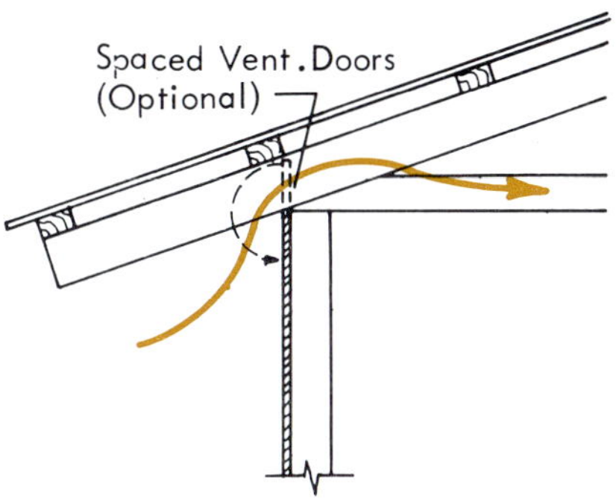

Mild Climate

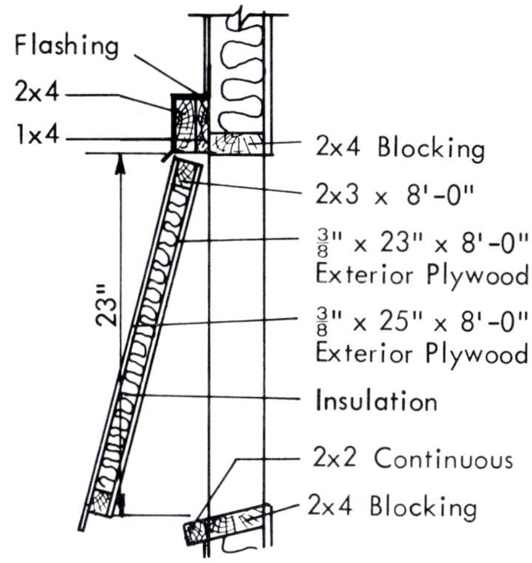

Summer Ventilation Door

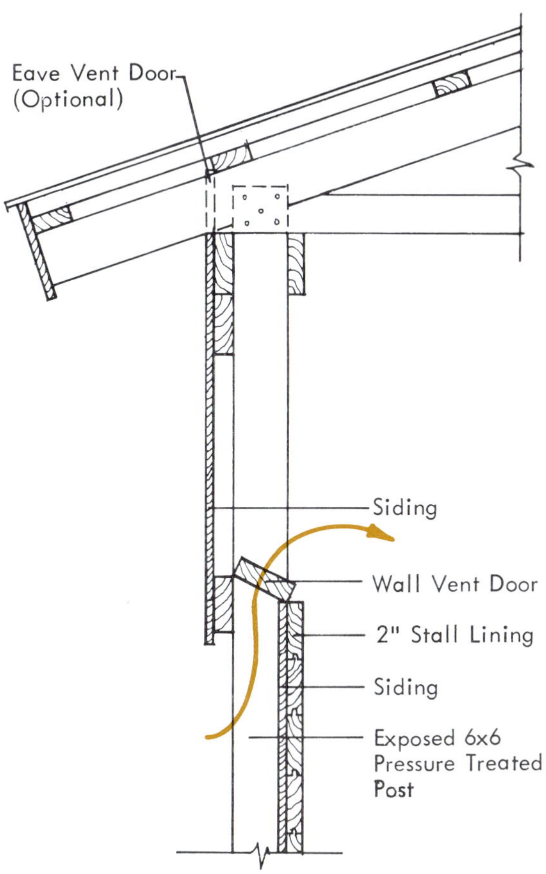

Alternate Back-Wall Vent

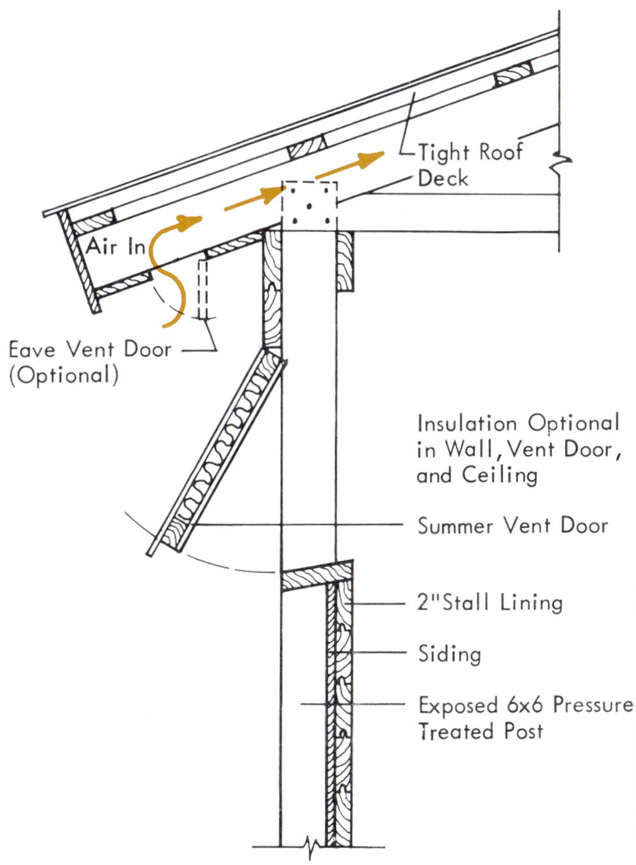

Wall and Eave Vent
Enclosed or Open Front Buildings

Ridge Vents

Provide a slot at the ridge—6" wide in cold zones, 4" wide in milder zones. Ridge ventilation can be provided by leaving off the ridgeroll with metal roofing. Protect exposed trusses with small metal covers.

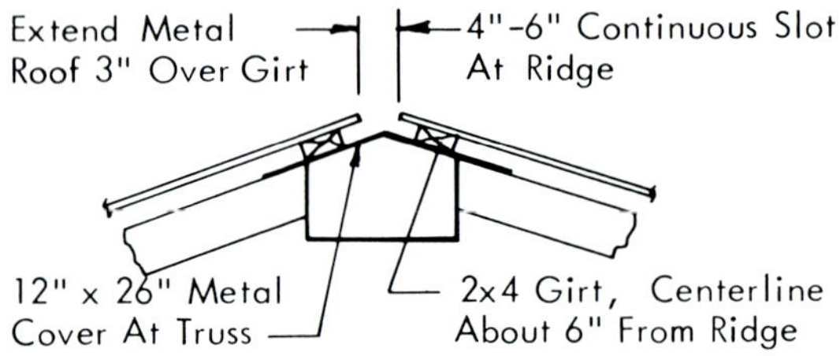

When appearance is a consideration, cupola-type ridge ventilators may be desirable.

COLD ENCLOSED HOUSING

In cold housing, if the building is not insulated, there is little difference between the temperature in the barn and the outdoor temperature. Freezing temperatures often occur in the barn before outside temperatures drop below 20°F. Temperatures can usually be maintained above freezing in an insulated barn at outside temperatures above 15°F without supplemental heat.

Cold housing is usually ventilated by natural air movement. This requires wall and roof openings and sometimes assistance from fans for moisture and temperature control in winter and summer.

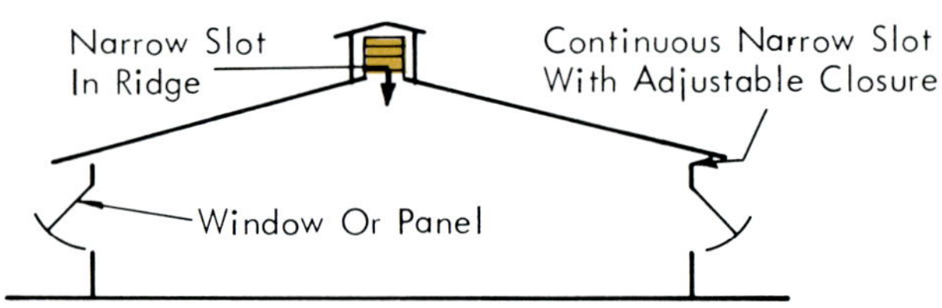

It is touch-and-go to depend on natural air movement for ventilation in an enclosed building with adjustable and fixed openings. The adjustable openings are manually controlled, consisting mainly of windows and wall panels, but including ridge ventilators and eave-slots equipped with adjustable dampers or closures. The fixed openings are often narrow slots under the eaves and slots or ventilators located at the ridge or high point of the roof.

Ventilating an enclosed building successfully by manipulating air inlets and outlets requires vigilance, skill, and luck on the part of the operator, as well as cooperation from the weather. Moisture problems in enclosed buildings that depend on natural air movement for ventilation are not uncommon, especially during calm humid weather. Timely and skillful adjustment of the air controls generally provides ventilation that is adequate. The use of circulation fans in the building improves hot weather animal comfort on calm days.

WARM ENCLOSED HOUSING

Warm housing refers to a tightly constructed and insulated building, equipped and managed to maintain inside temperatures 10°-15°F above freezing in the coldest weather regardless of building size, number of animals housed, and practices used by the operator to control moisture.

The following provisions for temperature and moisture control make warm housing possible:

- Insulation in the walls and ceiling.
- Vapor barriers to protect the insulation from moisture.
- Supplemental heat for temperature and moisture control.
- Fans to move air and to control moisture, odors, and temperature.
- Adjustable vents, doors, or other air inlets and outlets.

INSULATION

The amount of insulation needed in the walls and ceiling to slow down the flow of heat both into and out of the building depends on:

- Climate in the area.
- Type of insulation used.
- Type and size of the building.
- Number and size of the animals housed in the building.
- Amount of supplemental heat the operator is willing to furnish.

Table 6. Water vapor transmission of common building materials, perms[1]

1 Perm = 1 Grain/hr-ft^2-in.hg

Material	Perms
Vapor Barriers	
Aluminum foil (1 mil)	0.0
4 mil polyethylene (plastic film)	.08
Kraft and asphalt building paper	.3
Two coats of aluminum paint (in varnish) on wood	.3-.5
Common Building Materials	
1" extruded expanded polystyrene	1.2
15 lb tar felt building paper	4.0
1/2" insulation board, uncoated	50 to 90
1/4" 3-ply exterior plywood	.7
4" brick masonry	.8
4" poured concrete wall	.8
4" glazed tile masonry	.12
8" concrete block	2.4

[1]From ASHRAE Handbook of Fundamentals, 1967

Recommended Insulation

Climate Zone	Resistance (R)1 Walls	Ceilings
Mild	9	12
Moderate	12	16
Cold	14	23

[1]Total of resistances of insulation, lining and siding, surfaces, and air spaces.

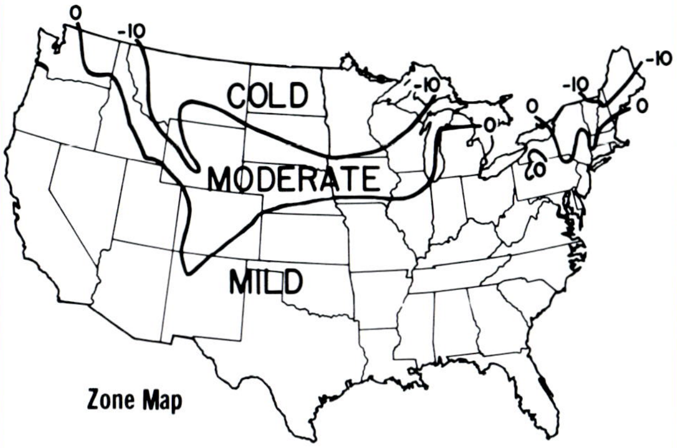

Table 7. Insulation values for common materials[1]

Material	Insulation value[2] Per inch thickness	For thickness listed
1. Batt or blanket insulation		
Glass wool, mineral wool or fiber glass	3.70	
wood fiber	4.00	
2. Fill-type insulation		
Glass or mineral wool	3.51	
Vermiculite (expanded)	2.13 to 2.27	
Shavings or sawdust	2.22	
Paper or pulp products	3.57	
3. Rigid insulation		
Wood fiber sheathing	2.27 to 2.63	
Expanded polystyrene, extruded	3.85	
Expanded polystyrene, molded	3.57	
Urethane foam (aged)	5.88	
Glass fiber	4.01	
4. Ordinary building materials		
Concrete, poured	0.08	
Plywood, 3/8″	1.25	.47
1/2″	1.25	.63
Hardboard, 1/4″	.72	.18
Cement asbestos board, 1/8″		.03
Lumber (fir, pine) 25/32″	1.25	.98
Wood beveled siding 1/2″ x 8″		.81
Asphalt shingles		.44
Wood shingles		.94
5. Window glass, includes surface conditions		
Single-glazed		.89
Single-glazed with storm windows		1.79
Double-pane insulating glass		1.5 to 1.75
6. Air space (3/4″ or larger)		.90
7. Surface conditions		
Inside surface		.68
Outside surface (15 mph wind)		.17
8. Reflective-type insulation		
Aluminum foil	Ceiling	Wall
2 air spaces	5.00	5.92
3 air spaces	7.14	9.11

[1]From ASHRAE Handbook of Fundamentals, 1967.
[2]Mean temperature of 75°F.

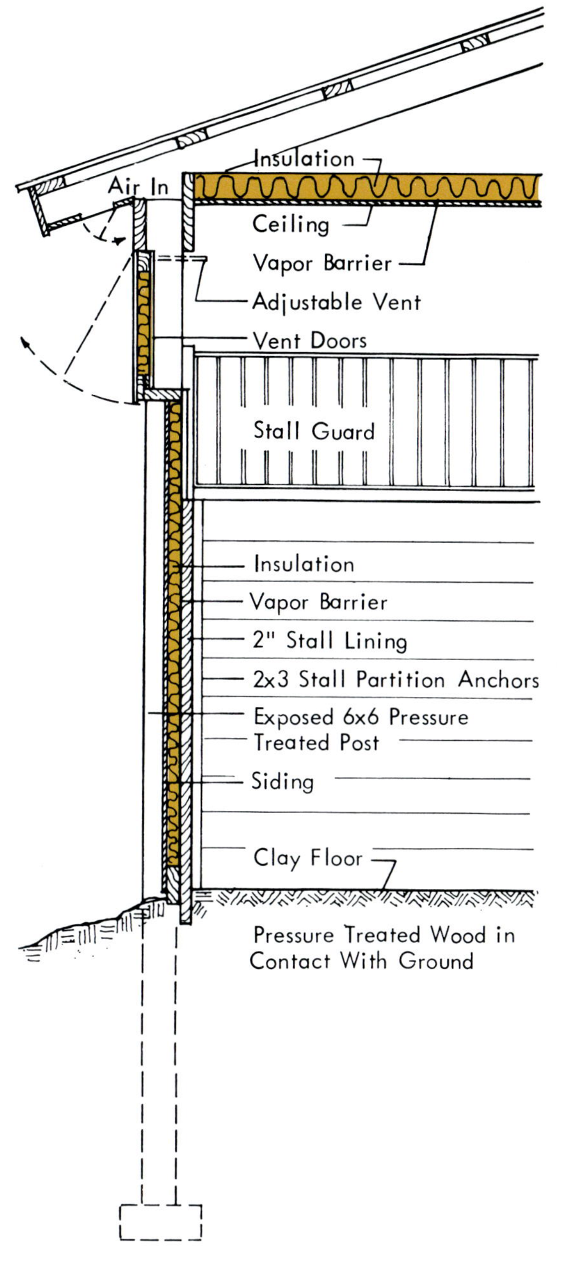

Typical insulation placement

VAPOR BARRIER

A vapor barrier is needed to prevent or slow down passage of water vapor through walls and ceilings. It must be installed between the insulation and the inside wall or ceiling lining. Vapor barriers to be effective must have a perm rating of 1 or less.

Many materials can be used as a vapor barrier; the most common is 4-mil polyethylene plastic. Such a membrane-type vapor barrier must be protected from mechanical damage by applying it under the finish of the inside wall or ceiling. Inside, rigid wall finishes, which are also vapor barriers, do not need this protection; however, joints or laps must be sealed with vapor-proof caulking compound or pressure-sensitive tape.

SUPPLEMENTAL HEAT

Heat must be supplied to warm housing to control moisture and maintain indoor temperatures above freezing when outside temperatures fall below about 15°F. The usual source of heat in a warm barn is the animals. Horses normally produce 200 to 300 Btu (British thermal units) per hour for each 100 lb of animal weight.

The design of most warm animal barns is based on an animal density of 15 to 40 lb/sq ft of total floor space in the barn. The density in a horse barn with two rows of 12' x 12' box stalls and a 12' center alley is as low as 2 lb/sq ft. Because of this low density, more heat is needed than the horses can provide. Consequently, the insulation requirement for a warm horse barn is greater than for other mature animals, and supplemental heat has to be supplied.

Stopping the ventilation fans to raise the temperature in the barn increases humidity, condensation, and odors—the barn becomes clammy and smelly.

The approximate amount of supplemental heat required for temperature and moisture control depends on:

- Climate of the region.
- Insulation rating of the walls and ceiling.
- Amount of heat and moisture produced by the animals housed.
- Size of the building.
- Ventilation rate (cfm/animal or air changes in the building/hr).
- Minimum acceptable temperature and humidity level inside the building in cold weather.

If the amount of insulation recommended on page 25, and the ventilation rates shown on page 28, are used for an all-stall barn, the approximate amount of supplemental heat needed for different conditions can be determined by adding the "building heat loss" and the "ventilation heat loss" for the selected minimum inside winter temperature. This simplified procedure assumes that animal heat is sufficient to control moisture. Additional heat will be needed if higher ventilation rates are used.

The "building heat loss" (Btu/hr) is determined by multiplying the inside wall and ceiling areas (sq ft) by the indicated multiplier for the zone and adding the results.

The "ventilation heat loss" (Btu/hr) is determined by multiplying the weight of the animals by the multiplier for the appropriate climate zone. Ventilation heat loss will increase about another 10% for each 5 cfm increase in ventilation rates above those recommended on page 28.

Computing Supplemental Heat

Building Heat Loss (Btu/hr)

Select the Ceiling and Wall multipliers from Table 8 for your desired inside conditions and your climate area (map, page 25).

Building heat loss = ceiling area (sq ft) x Ceiling multiplier + wall area (sq ft) x Wall multiplier

Example: 36' x 36' barn with 5 box stalls is to be kept at 45°-50°F in Indiana.

Ceiling area = 36'x36' = 1296 sq ft
Ceiling multiplier (mild zone) = 4.2
Wall area = 4 walls x 8' high x 36' long = 1152 sq ft
Wall multiplier (mild zone) = 5.5

Building Heat Loss = 4.2 x 1296 + 5.5 x 1152 = 5443 + 6336 = 11,779 Btu/hr.

Table 8. Heat loss multipliers

	35°F inside			45°-50°F inside		
	cold	moderate	mild	cold	moderate	mild
Ceiling	2.4	2.8	2.9	3.0	3.7	4.2
Wall	3.9	3.8	3.9	5.0	5.0	5.5
Ventilation	3.0	2.4	1.9	3.9	3.2	2.7

Ventilation Heat Loss

Select the Ventilation multiplier for your inside temperature and climate zone.

Ventilation Heat Loss = Weight of horses (lb) x multiplier

Example: Assume 5, 1000 lb horses in the barn of the example above.
Ventilation Heat Loss = 5 x 1000 lb x 2.7 = 13,500 Btu/hr

Total Heat Loss = Building Loss + Ventilation Loss

Example: Supplemental Heat needed = Total Heat Loss = 11,799 + 13,500 = 25,299 Btu/hr

NOTE: When figuring areas (sq ft) and the roof is the ceiling, treat the roof area as the ceiling area and the ends of the building as part of the total wall area.

Install heating equipment according to the manufacturers' recommendations.

IMPORTANT: Heaters using gas, liquid, or solid fuel must be vented to the outside by means of a fire-safe chimney or flue, to carry the by-products of combustion out of the building.

Set the thermostat to shut the heater off below the temperature at which the main winter fans start to operate (usually 50°-55°F).

VENTILATION SYSTEMS

There are two ventilation systems in common use for warm housing—pressure ventilation and exhaust ventilation. Either system will adequately control air quality if it is designed, installed, and operated properly.

Pressure ventilation systems use fans to force fresh air into a building and let old air flow naturally out through vents. For buildings 24'-36' wide install the fan duct or baffles in the center. For wider buildings space the fan duct or baffles about 20' apart.

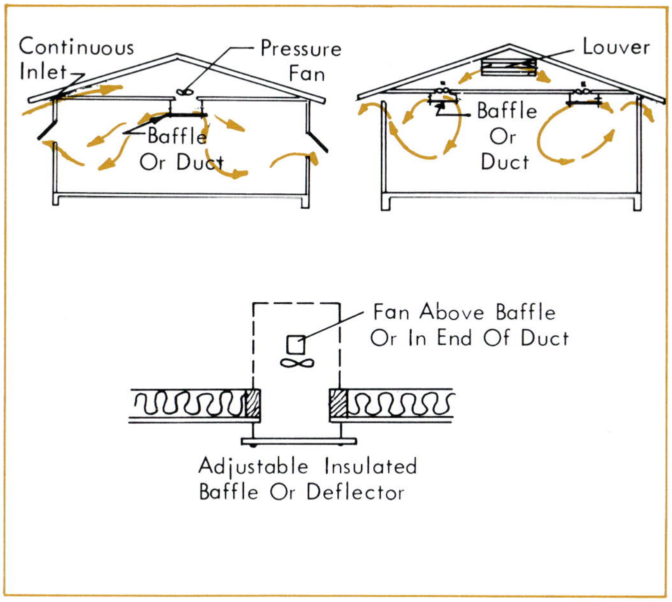

Pressure ventilation systems

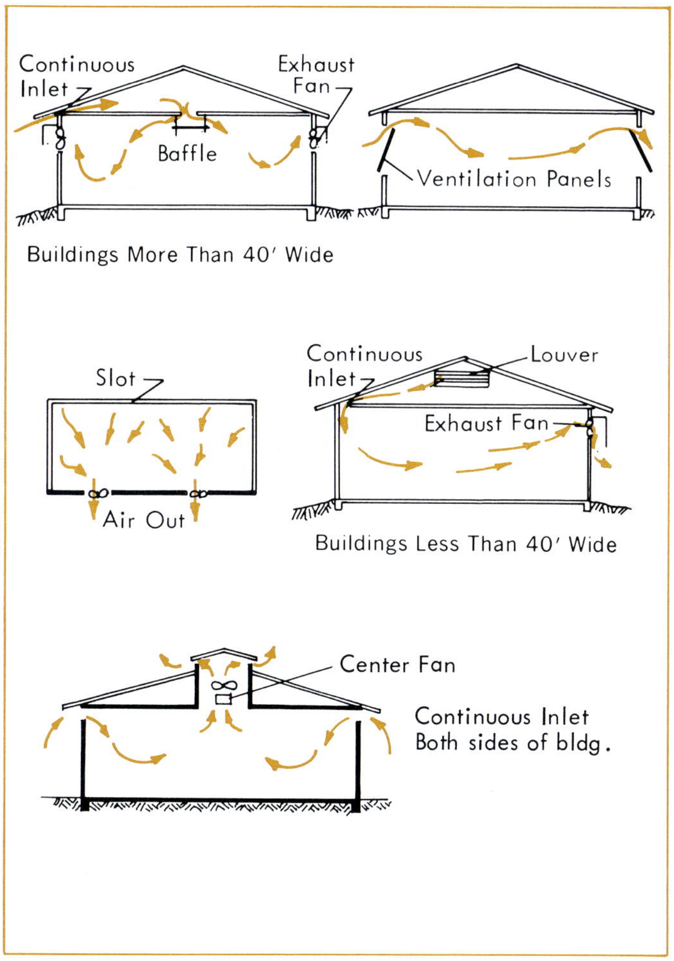

Exhaust ventilation systems

Exhaust ventilation systems use fans to force the old air out of a building and let the new air come in naturally through the vents or openings.

Use wall fans and continuous adjustable slot inlets at the top of the walls for buildings less than 40' wide.

Use wall fans and adjustable center diffusing inlets when the building is 40'-50' wide. Install two rows of inlets for widths over 50', or use center exhaust fans with slot inlets at the top of the outside walls.

Open large wall panels and doors for natural ventilation during hot weather or increase fan capacity.

Both pressure and exhaust ventilation systems require one or more fans that move different amounts of air. Some fans must run at all times, while others may be needed for only short periods. The needed ventilation rate varies with the outdoor air temperature and the size and number of animals. With cold outdoor temperatures, little ventilation is needed, but as outdoor temperatures increase, more ventilation is needed.

FANS

Location

Locate the fans in tight walls or in the ceiling away from large openings such as doors, windows, stair wells and hay chutes. Close large openings in cold weather for forced air ventilation. Face exhaust wall fans away from prevailing winds when possible. Fans may be grouped together or spaced apart as design dictates.

Selection And Operation

A ventilating fan capacity of 100 cfm/1000 lb of animal weight should provide reasonable ventilation and temperature control for temperatures below 60°F outside. During extremely cold weather only about ¼ of the fan capacity is needed. Select one fan with one-fourth the needed capacity to run continuously, and control the other fan or fans with a thermostat. The one-fourth fan capacity may be obtained with a large fan and a variable speed fan control. During extended cold periods without supplemental heat it may be necessary to operate the fan manually to supply fresh air to the stable.

Provide some air movement at all times. Fan capacities are given in cfm, which means cubic feet per minute.

Install fans that have been rated at 1/8" static pressure. Select fan motors that are totally enclosed and have an overload protection device to prevent motor burnout or fire.

Use automatic shutters to prevent back drafts when the fans are not running and a wire fan guard to protect people, animals, birds, and the fan.

Locate the thermostat in the approximate center of the building at eye level. Set the thermostat at 45°F and adjust up or down as needed. Keep the thermostat out of direct sunlight and away from other heat sources.

In warm weather, doors or windows can be opened. In hot weather additional fan capacity may be needed to change the air in the building more often. Also, the use of circulation fans in the building will improve hot weather ventilation on calm days.

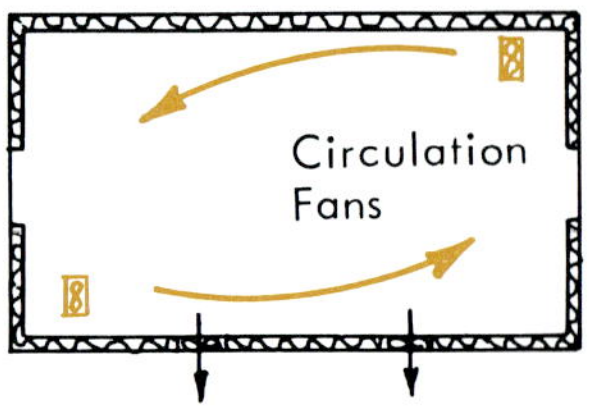

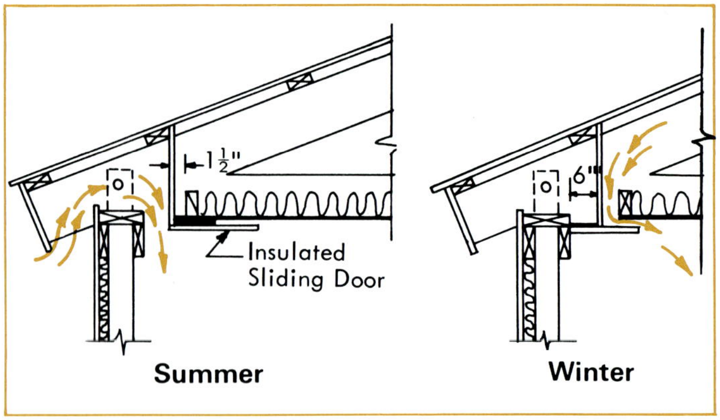

A slot inlet for fresh air

Air Inlets and Outlets

Distribution and air flow within a building is almost completely determined by location and design of air inlets (exhaust system) or outlets (pressure system) and not by the fan location.

Inlets, outlets, and other ventilation openings are needed to complete the movement of air through a building. They are used in a pressure system to release old air, and are used in an exhaust system to supply fresh air.

The number and size of openings needed depends on the fan capacity being used. As fan capacity is being increased, the size of openings must also increase.

The style and placement of openings are often determined by the ventilation systems used, and the placement of the fans.

Recommendations

Provide 1 sq ft of inlet or vent area for each 750 cfm fan capacity.

In barns with mows or attics, install inlets that allow incoming winter air to be tempered above the ceiling before being drawn or forced into the building. Close the inlets during hot weather to prevent hot attic air from being drawn into the building.

Provide an adjustable control over the inlets or outlets to vary the air flow. The outlets in a pressure ventilation system are usually self-closing doors.

The smaller fan capacities needed during cold weather require little air from the adjustable slot to balance the air movement. If the slot opening is not reduced, excess incoming cold air may cause drafts and reduce inside temperature.

Permit air to circulate through the space above the ceiling during summer months to prevent excessive heat buildup. Do not draw hot attic air into the animal shelter.

Do not install inlets where air from manure storage may be drawn in.

Provide large insulated wall panels that can be opened for natural ventilation during hot weather.

NOTE: Warm buildings are designed on the basis that they (1) have a fairly constant animal load and (2) that the barn remains closed most of the time. If either of these conditions is violated frequently, then operation of the heating and ventilation system is greatly impaired, and the barn will not function properly.

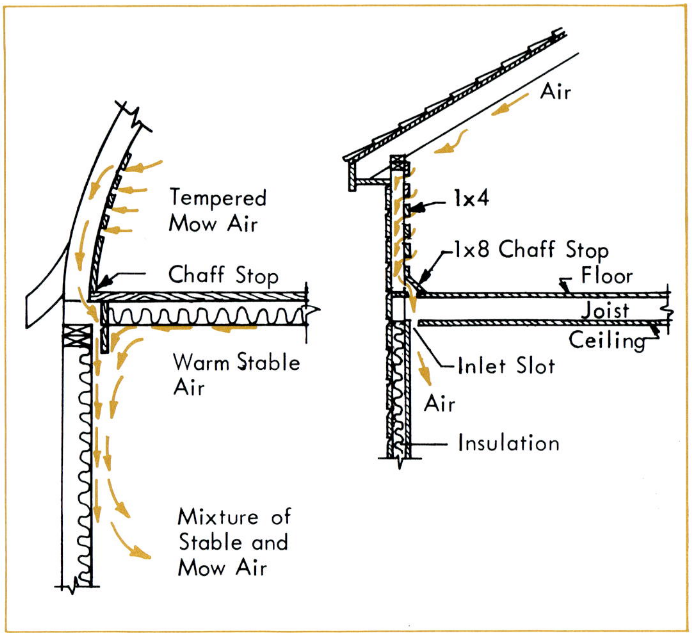

Continuous slot inlets for barns and hay mows

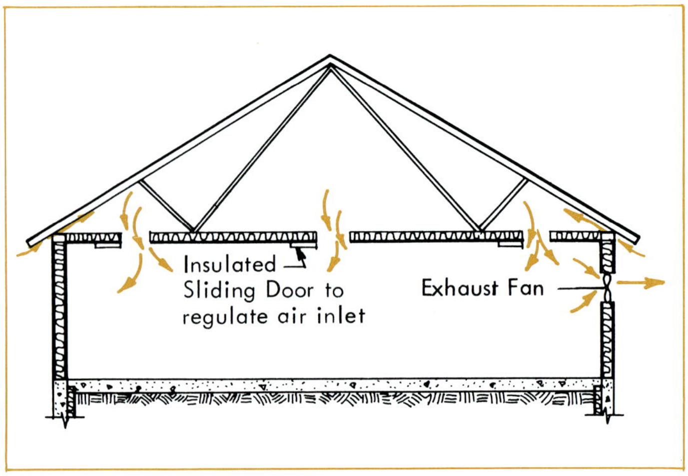

Fresh air inlets located in ceiling

CONSTRUCTION

TACK ROOM EQUIPMENT

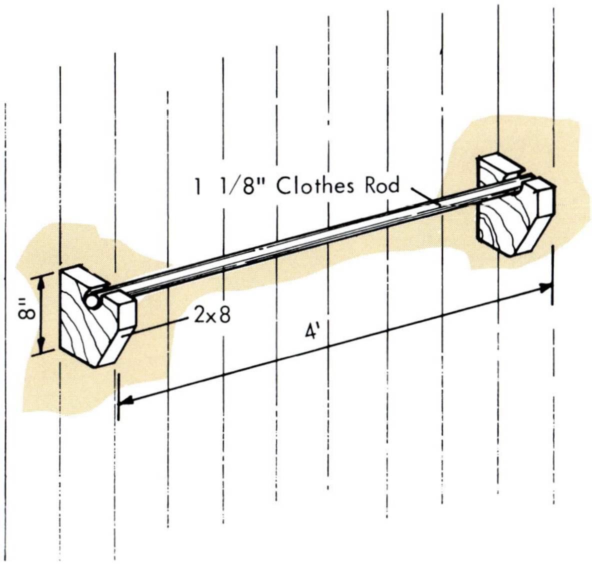

Blanket Rack

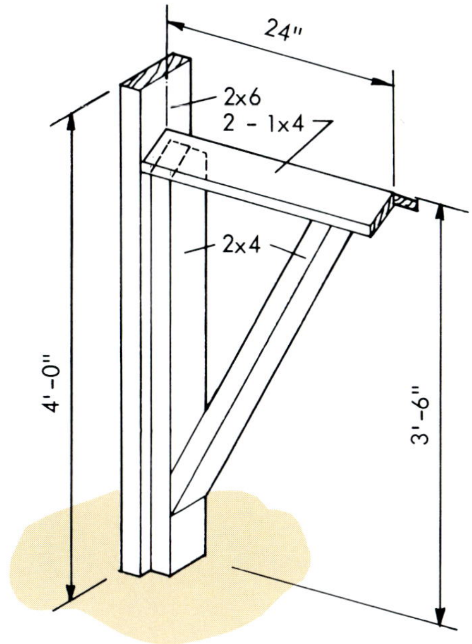

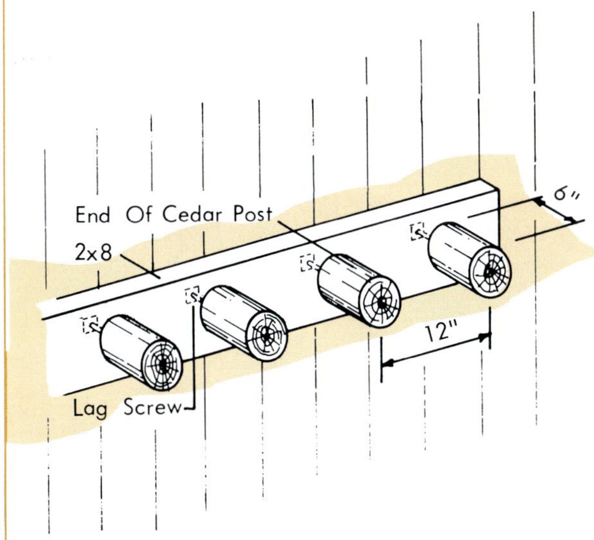

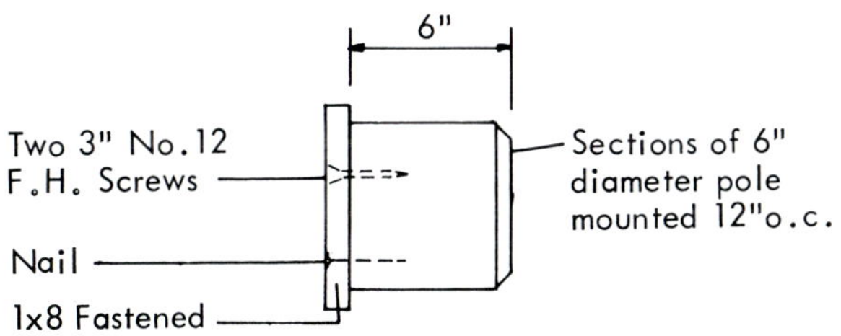

Option

Bridle Racks

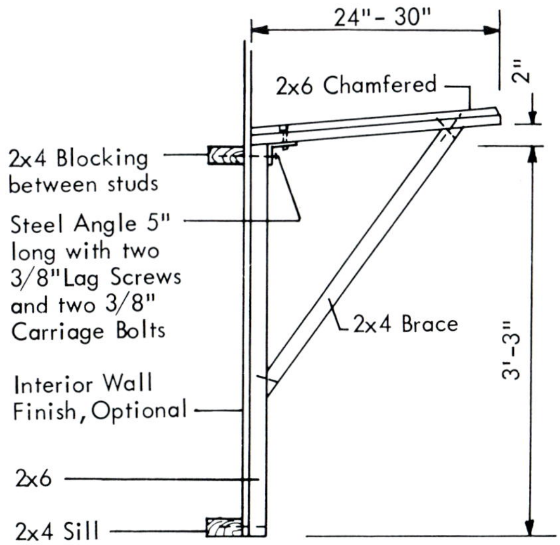

Space racks 24"to 30"apart

Saddle Racks

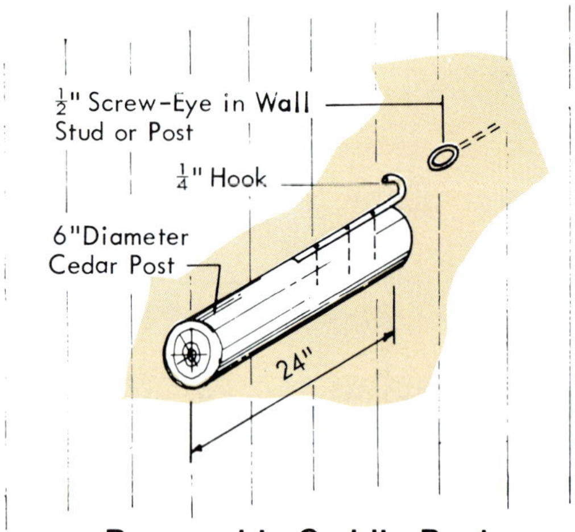

Removable Saddle Rack

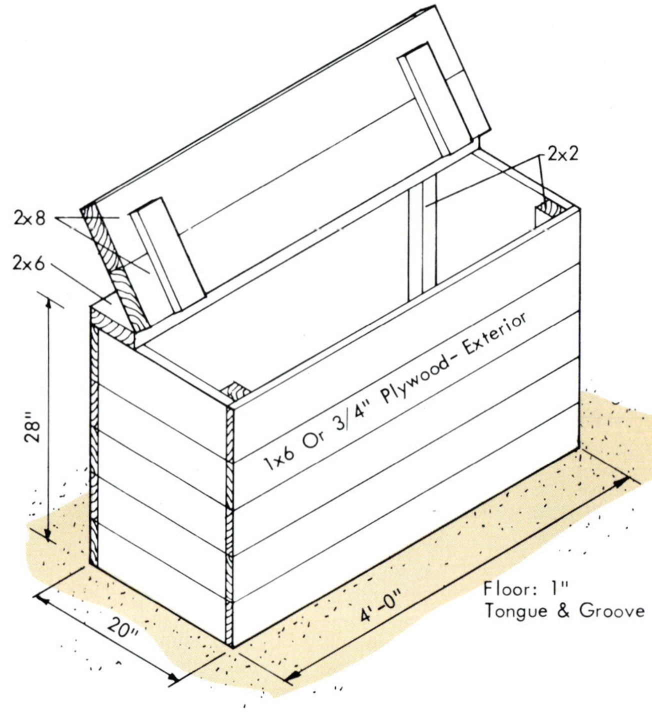

Feed Box With Bench Top

Cutting List

Item	No.	Description
A	4	2x4x4′-0″
B	2	1x4x18″
C	2	See Detail
D	2	1x4x5′-0″
E	2	1x4x30″
F	2	1x4x5′-0″
G	2	1x4x16″
H	2	1x4x14″
I	2	1x4x30″
J	4	1″x16″ Leather Straps
K	1	2x2x30″
L	1	¾″x12½″x5′-0″ Plywood
M	1	¾″x14½″x5′-0″ Plywood

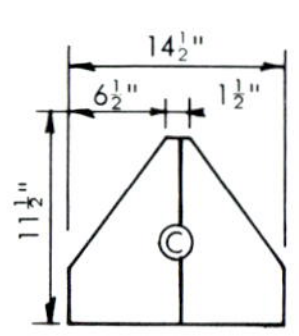

2-,2x8

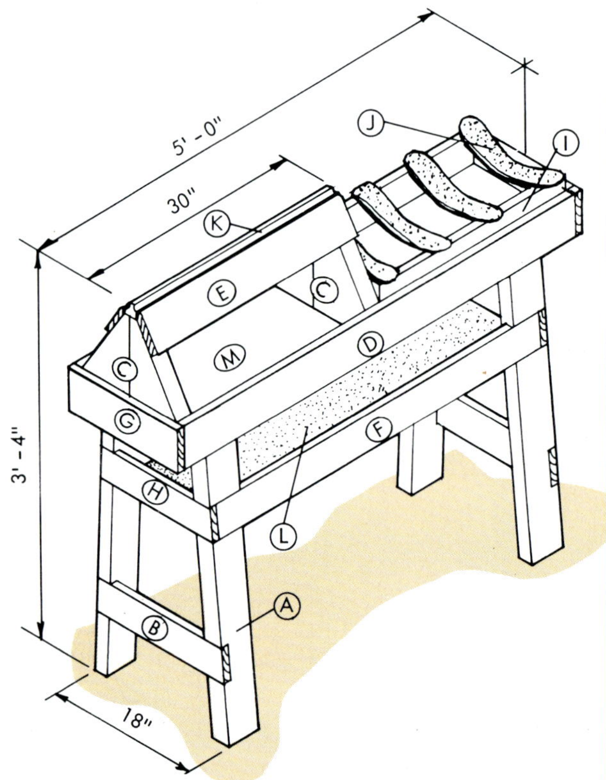

Saddle Cleaning Rack

Fences

Traditionally, fences for horses are made of wood. Wood is still the most widely used material for fencing roadways, paddocks, exercise lots, and areas used for special and competitive events. However, there are other satisfactory materials such as pipe and 4" diamond woven wire fence. When wire fencing is used, select a mesh that a colt cannot get his feet through. Use a protective board on top of wire fences. Use pressure-treated 5" diameter wood posts, or equivalent; space 8' on center and set 3' in the ground. Pipe fences may be attached to metal posts set in concrete.

Wood fences are normally made of 1x6 or 1x8 rough-sawed native lumber. The bottom board is set 9"-12" above the ground, and the other boards are spaced about 8" apart to obtain the desired fence height. If a higher and more rugged fence is desired, use 6" diameter posts and 2" thick fence boards.

The fence should be free of sharp projections, such as nails, bolts, and latches, which could in-ure an animal. No barbed wire, please! Planks may be pressure-treated, sprayed, or brushed with 5% pentachlorophenol (penta) in light oil for durability. For appearance, use a penetrating oil stain or quality lead-free paint.

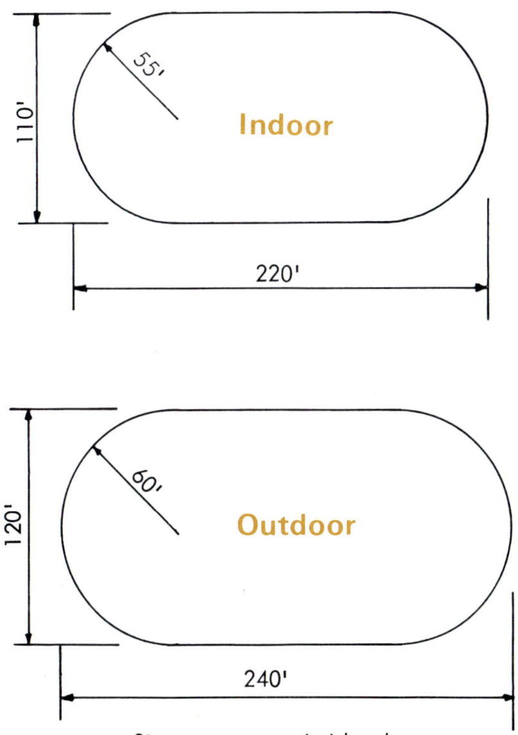

Sizes recommended by the National Horse Show Association

Horse Show Rings

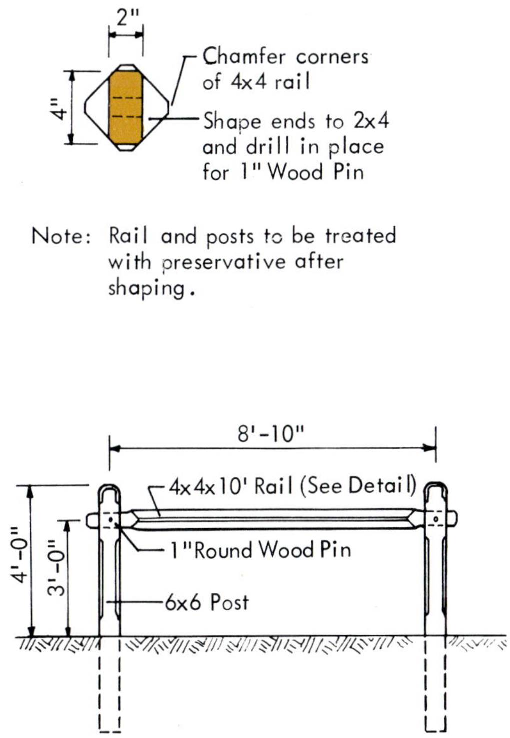

Hitching Rail

FENCES

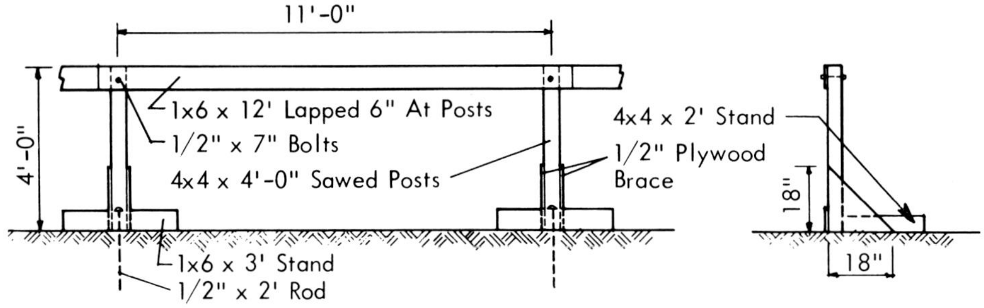

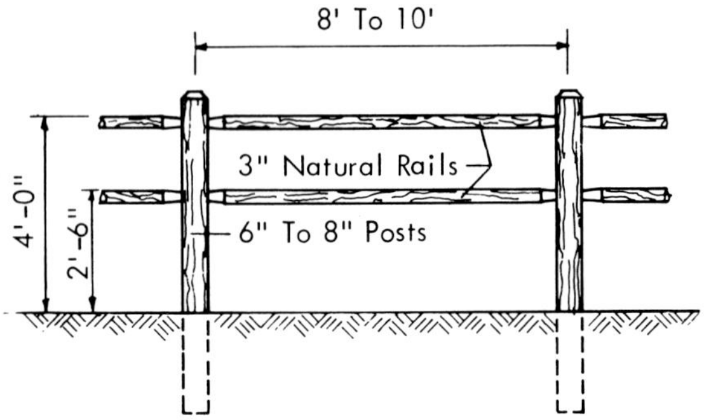

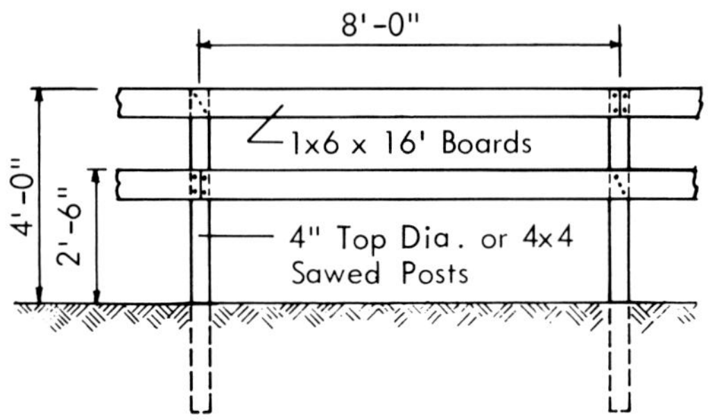

Show Ring Fences

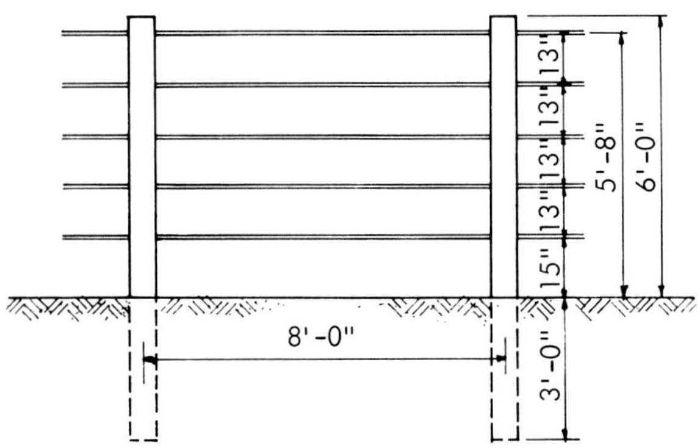

Cable

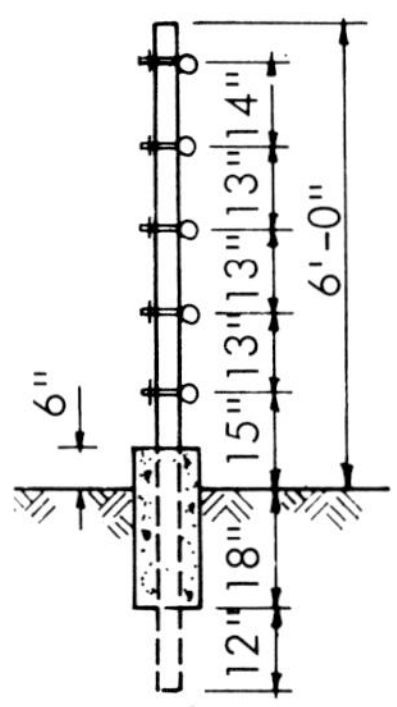

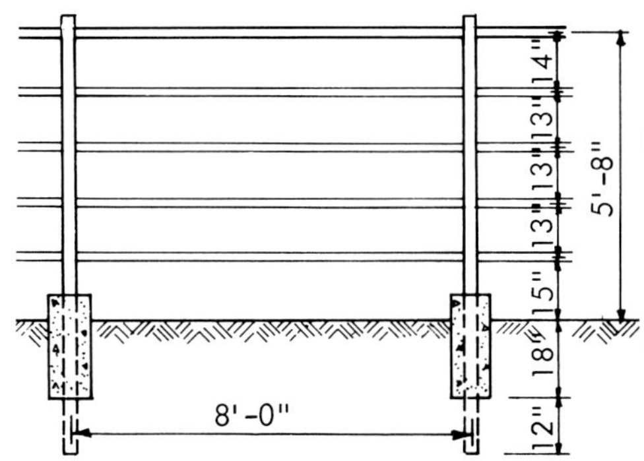

Pipe

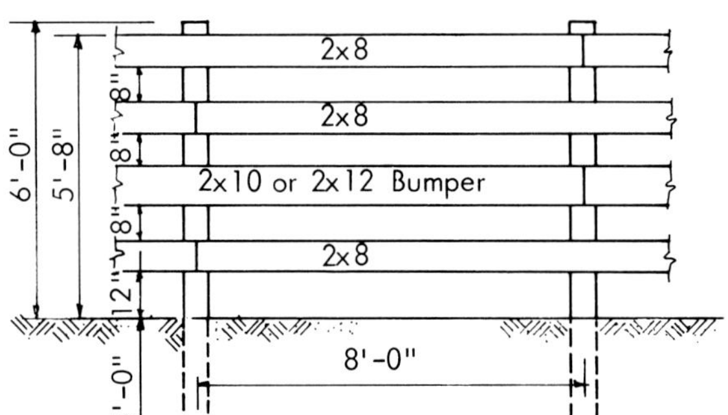

Plank

Line Fences

GATES

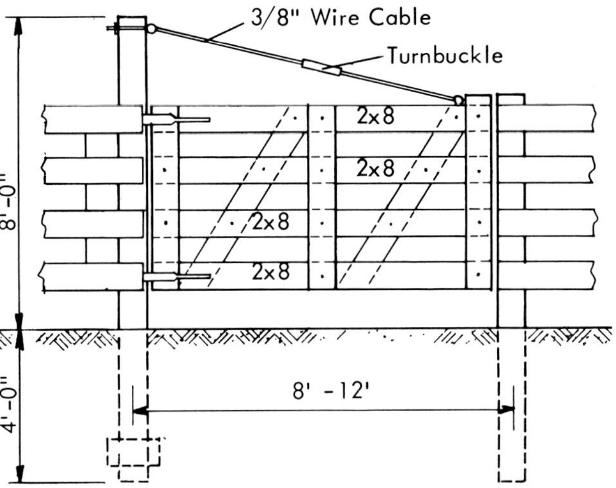

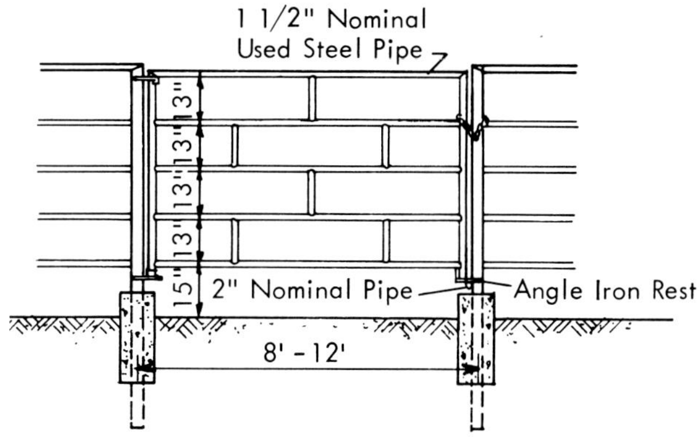

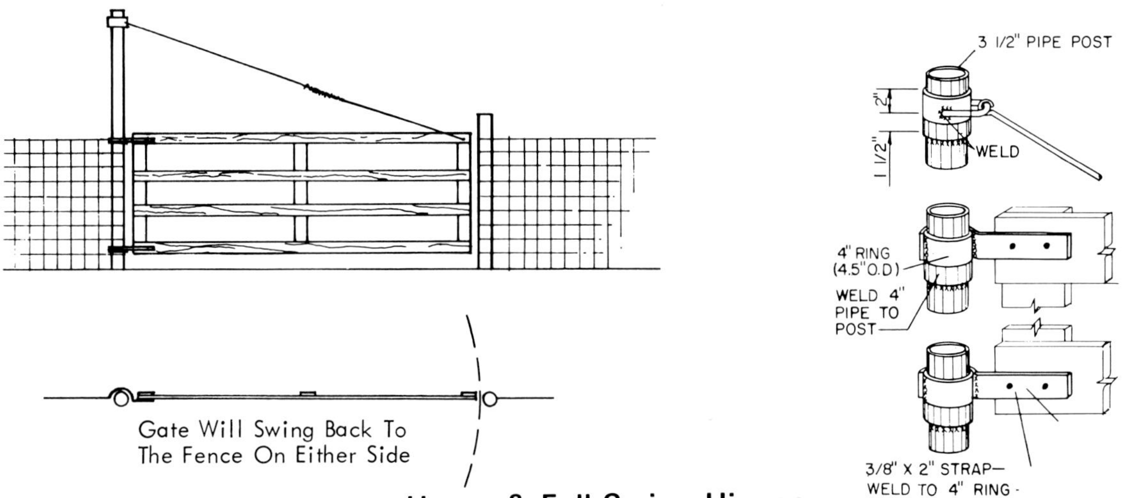

Heavy & Full-Swing Hinges

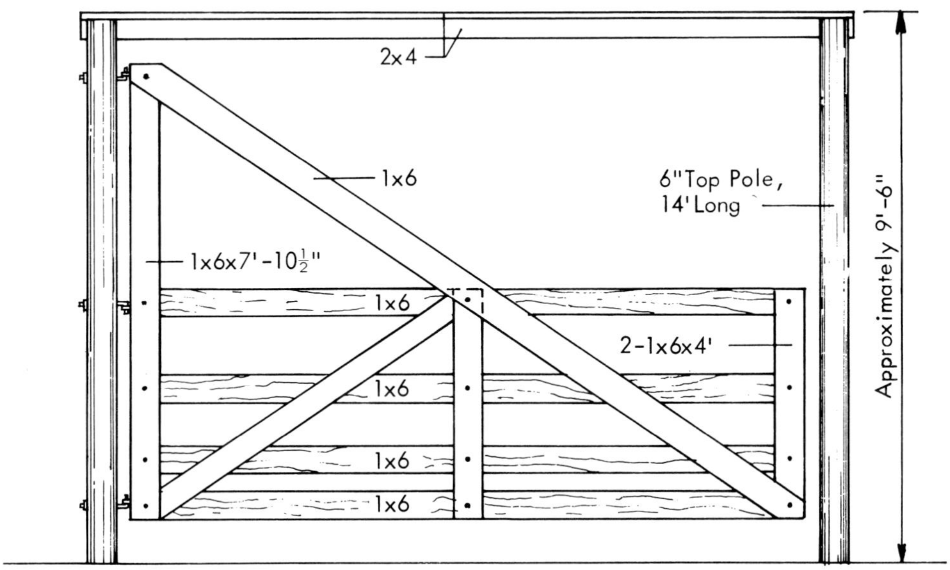

12' To 16' Gate

AUTOMATIC GATE

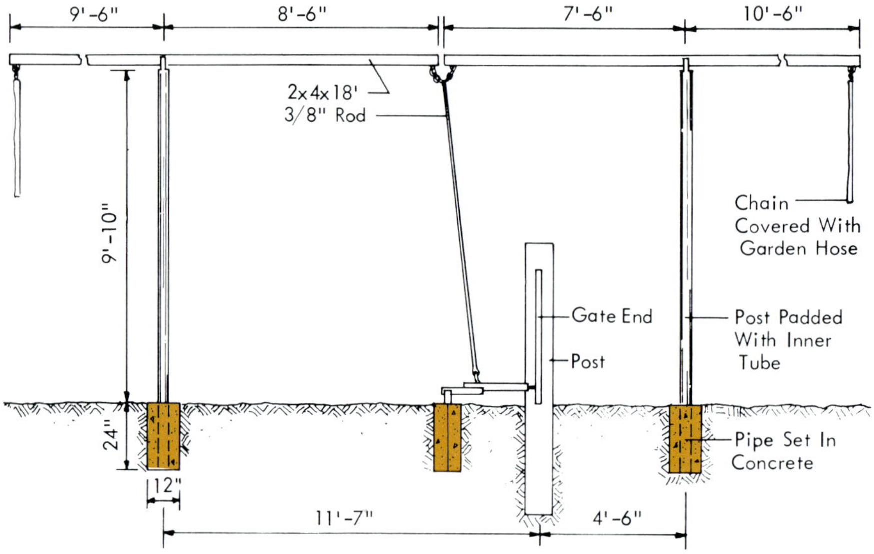

Elevation

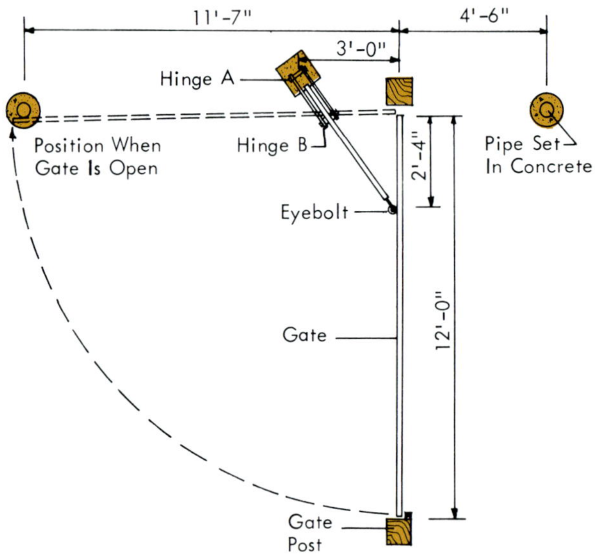

Plan

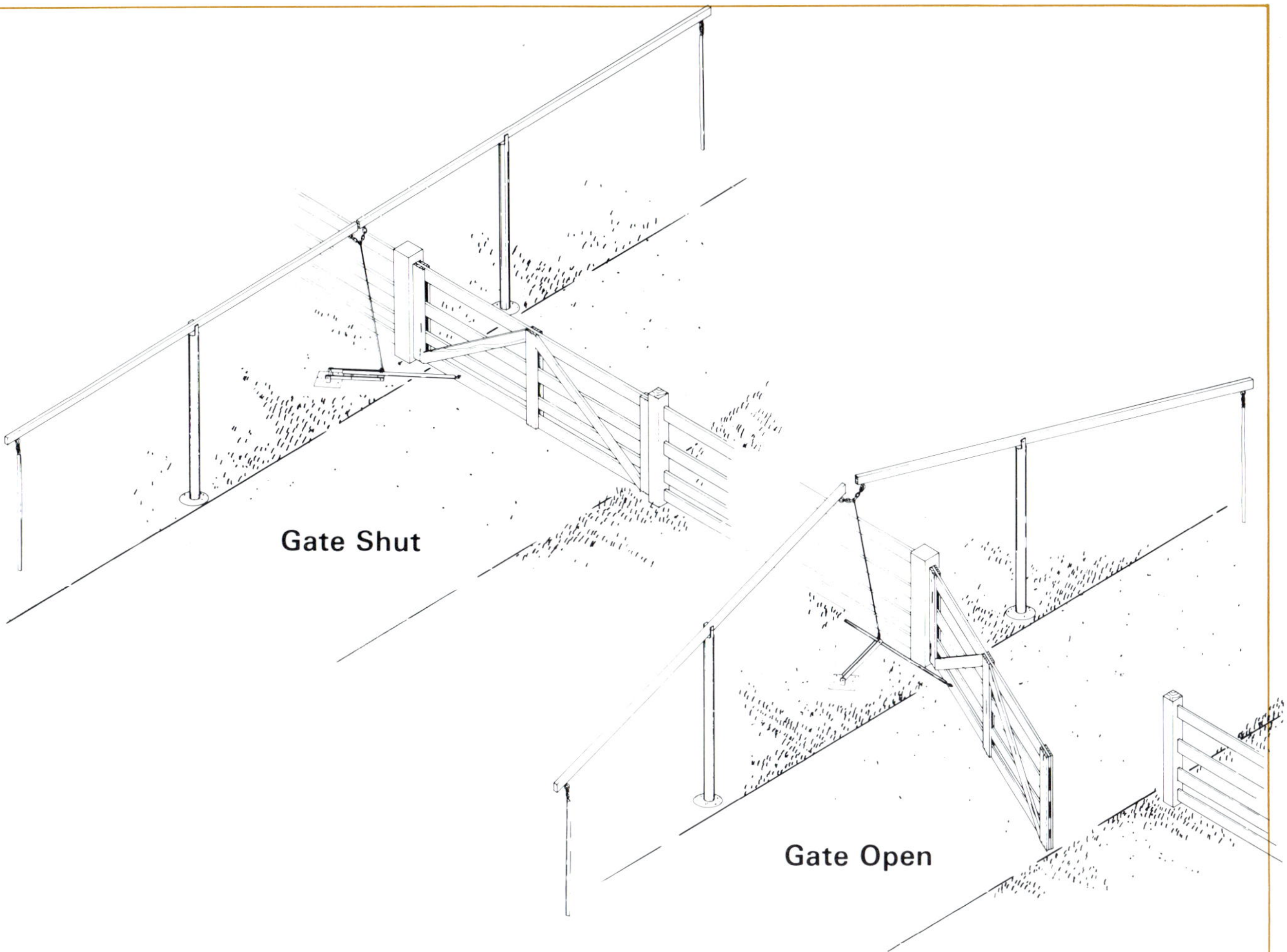

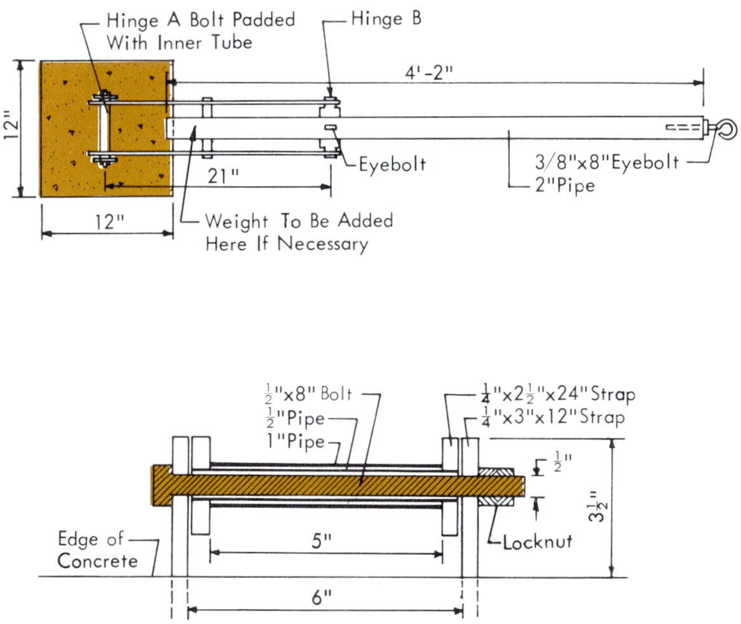

Hinge A Detail

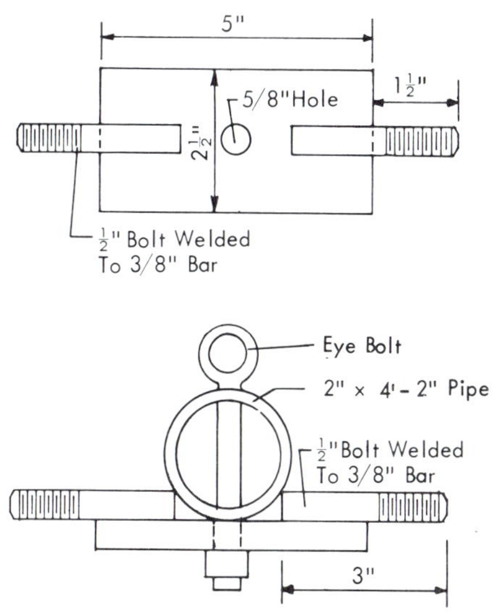

Hinge B Details

HINGES

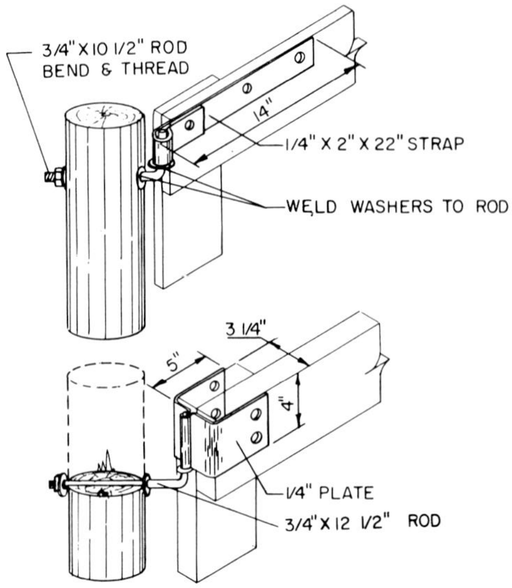

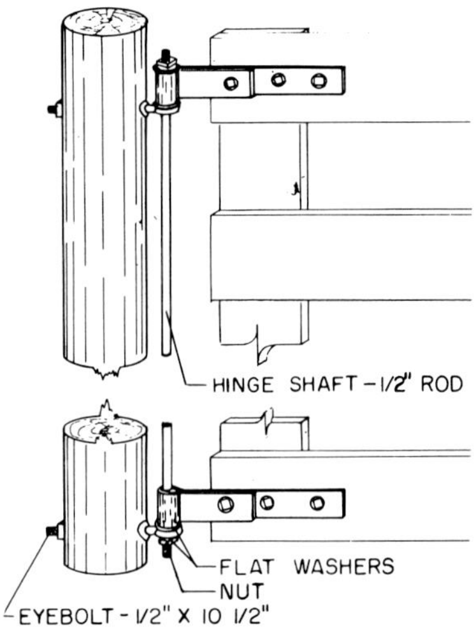

LOT & FIELD HINGE

LATCHES

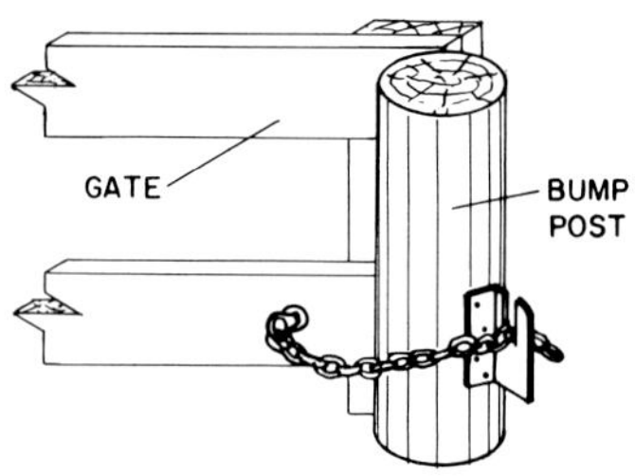

Hook & Chain

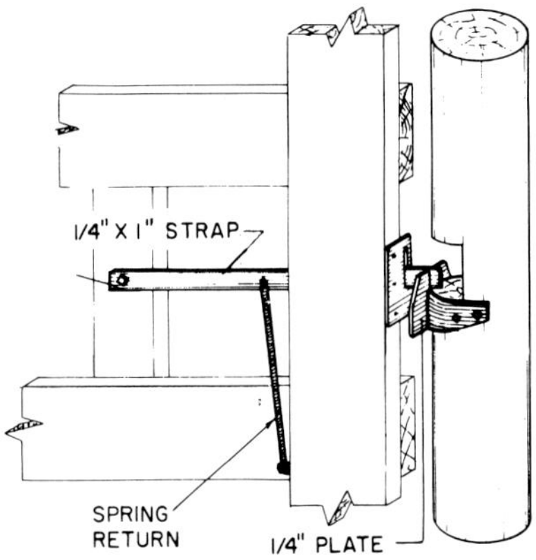

Strap Iron Latch
(Locks Automatically)

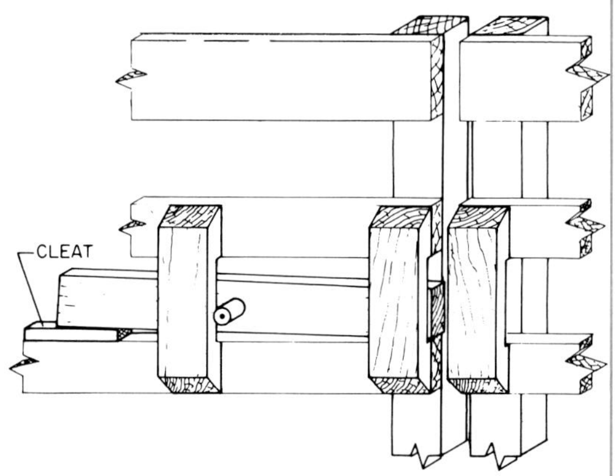

Slide Stick Latch

STILES AND PASSES

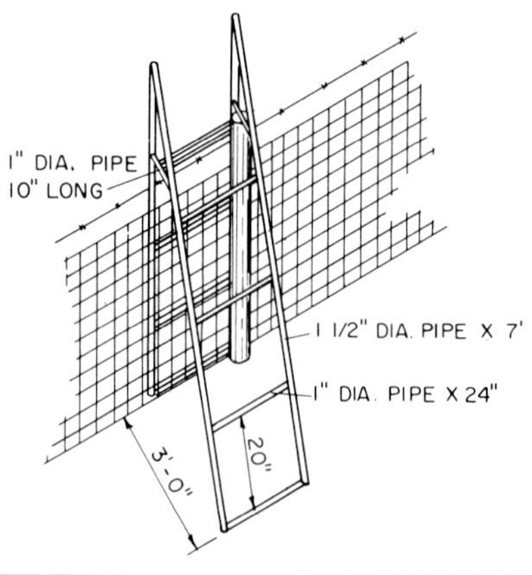

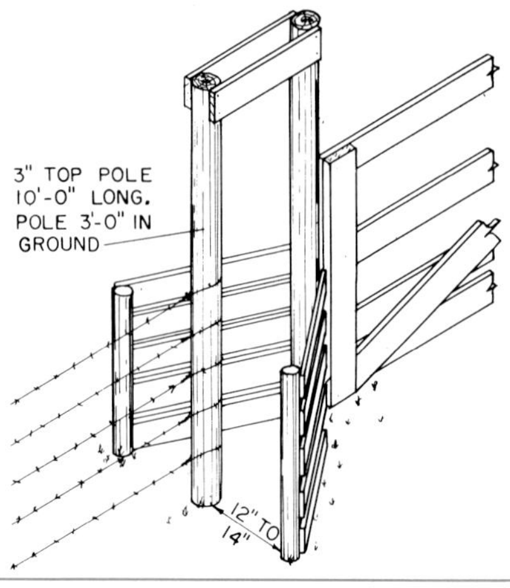

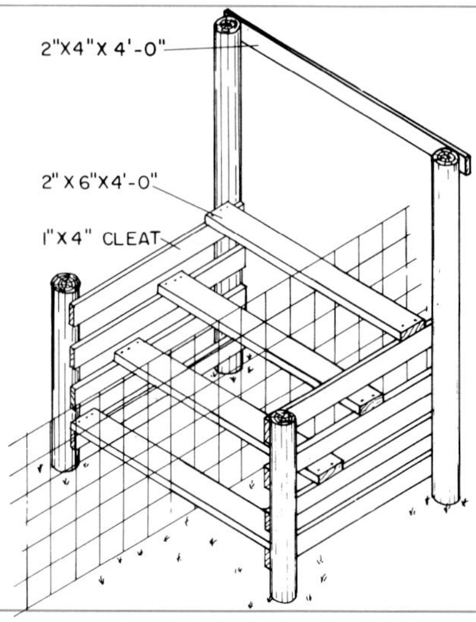

WINDBREAK FENCES

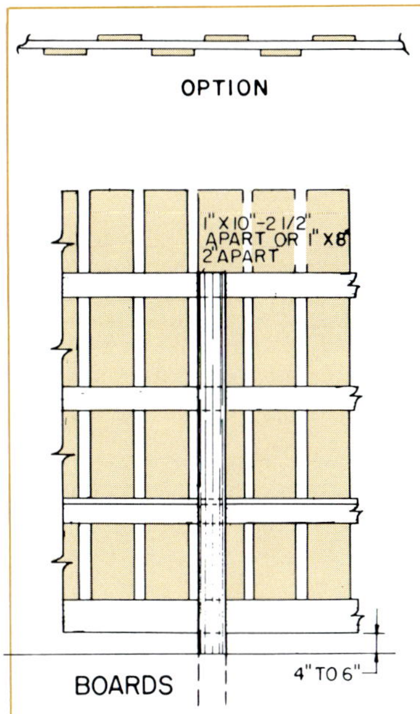

80% Solid

Better For Snow Protection Than Solid Fence

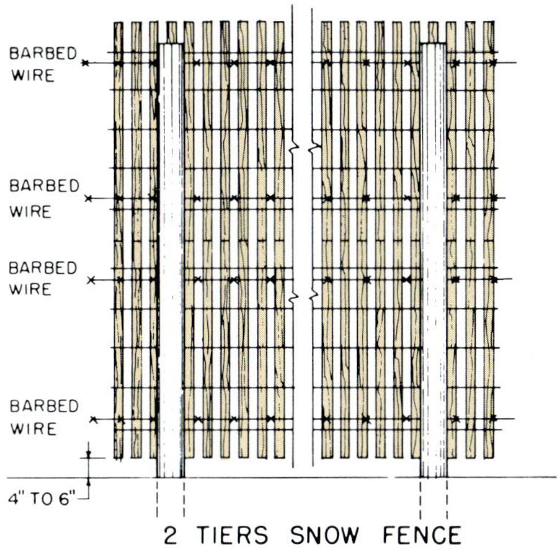

43% Solid

Plank Fence Splices

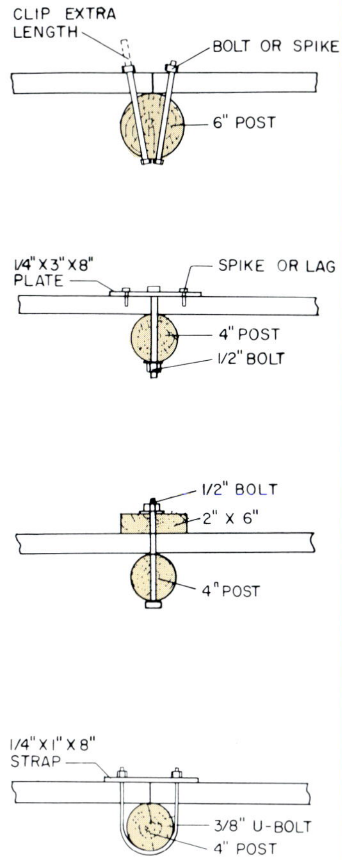

Wood Post Splice

Round Or Square Posts

POLES - 6" Top Diam., 10' O.C.

6' Fence	8' Poles	3'-6" Deep	3 Girts
8' "	10' "	3'-6" "	4 "
10' "	12' "	4'-0" "	5 "
12' "	14' "	4'-0" "	6 "
8' Snow Fence	12' "	4'-0" "	

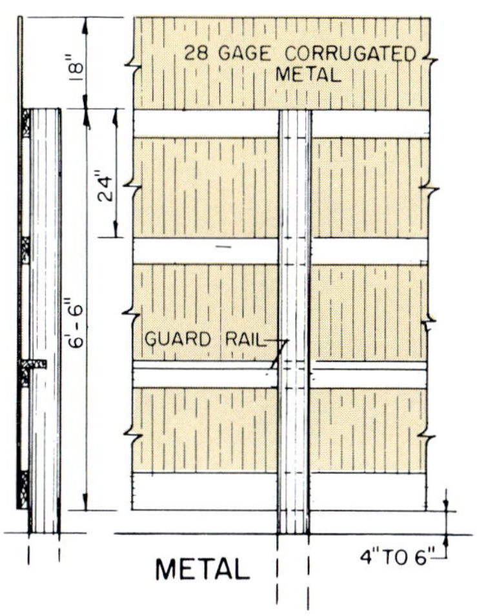

Solid

Better Wind Than Snow Protection

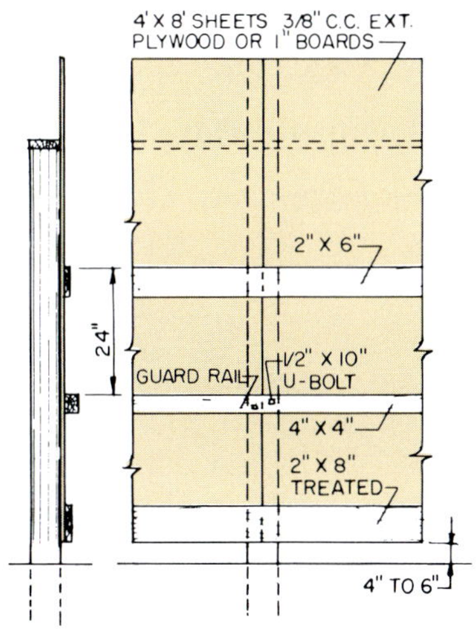

Plywood Or 1'' Boards

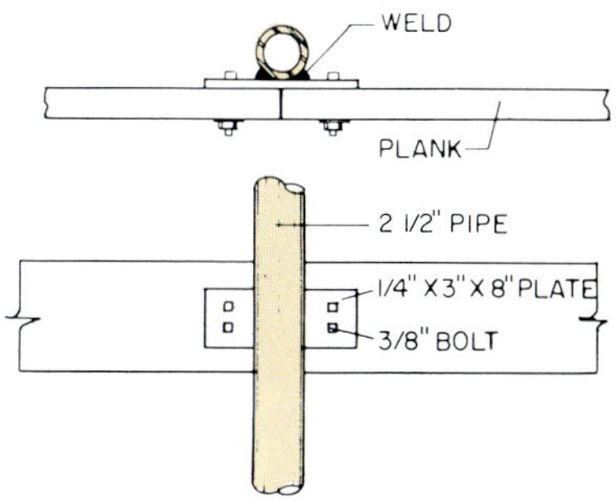

Pipe Splice Post

JUMPS

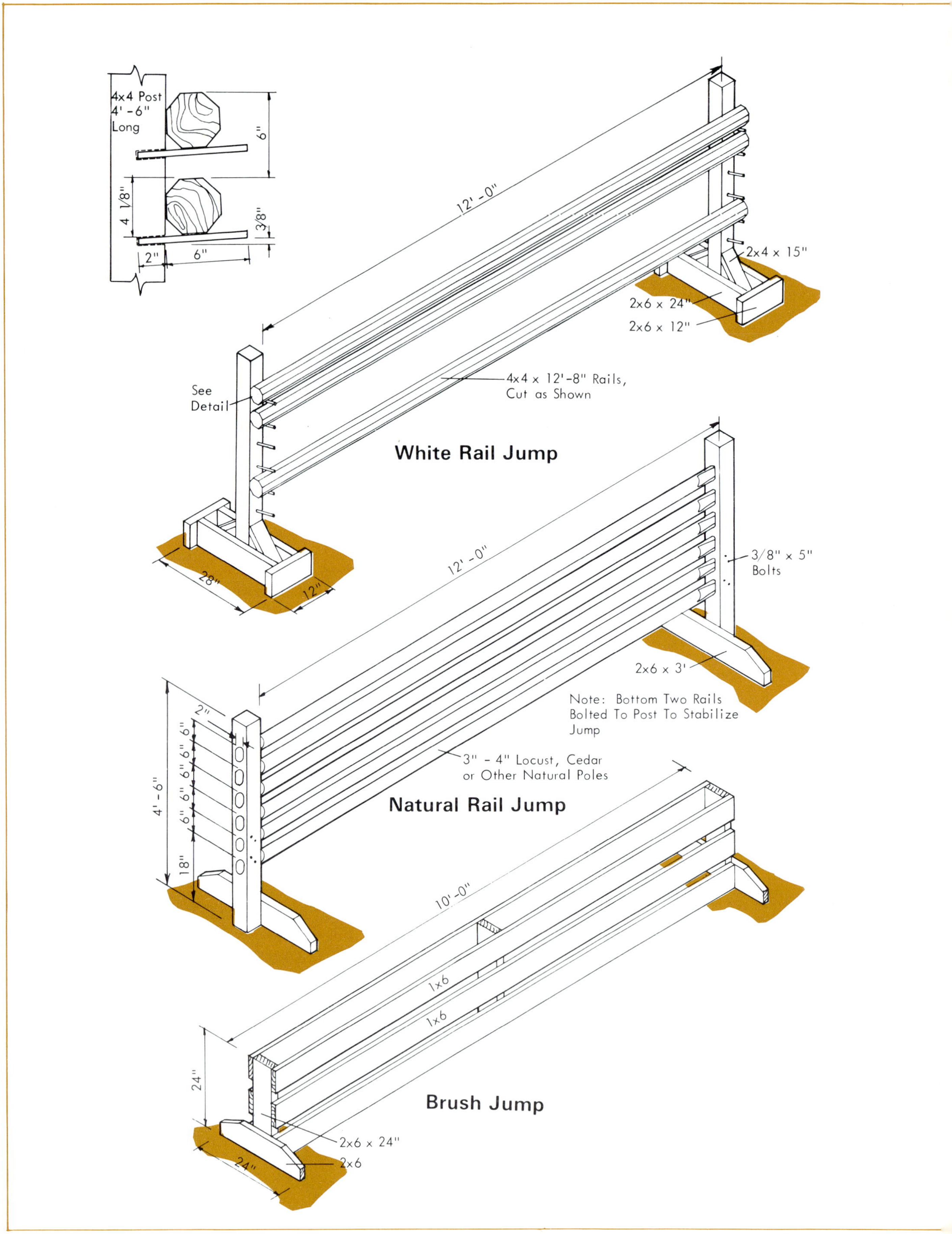
4x4 Post
4'-6"
Long
6"
4 1/8"
3/8"
2"
6"
12'-0"
2x4 x 15"
2x6 x 24"
2x6 x 12"
See
Detail
4x4 x 12'-8" Rails,
Cut as Shown
White Rail Jump
28"
12"
12'-0"
3/8" x 5"
Bolts
2x6 x 3'
Note: Bottom Two Rails
Bolted To Post To Stabilize
Jump
2"
6"
6"
6"
6"
6"
4'-6"
18"
3" - 4" Locust, Cedar
or Other Natural Poles
Natural Rail Jump
10'-0"
1x6
1x6
24"
Brush Jump
2x6 x 24"
2x6
24"

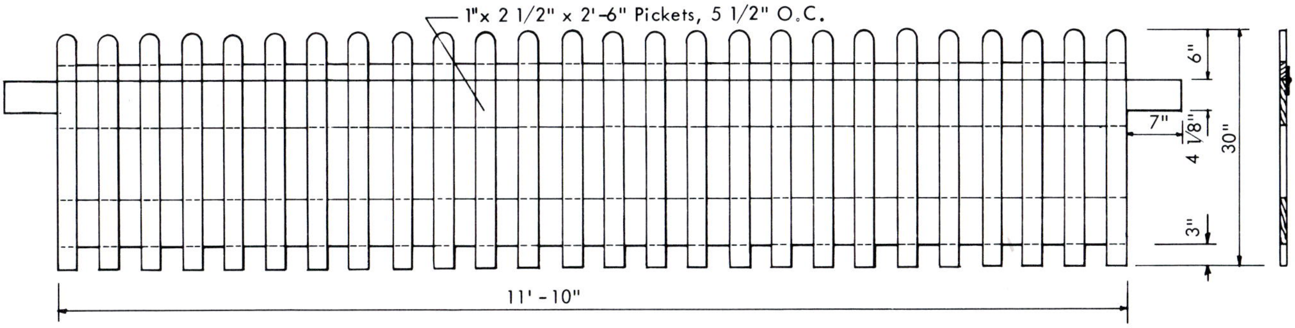

Picket Jump

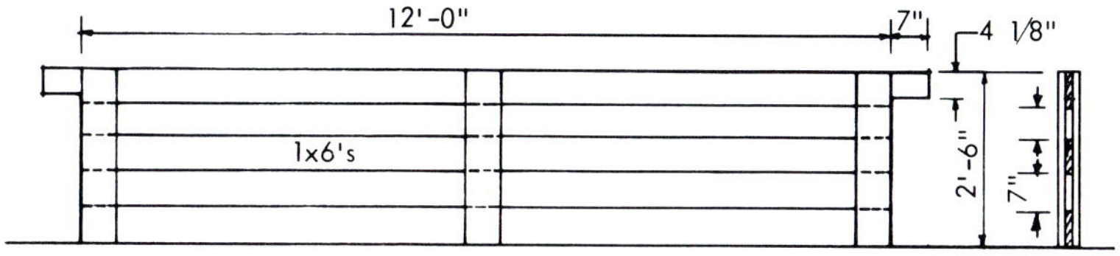

Plank

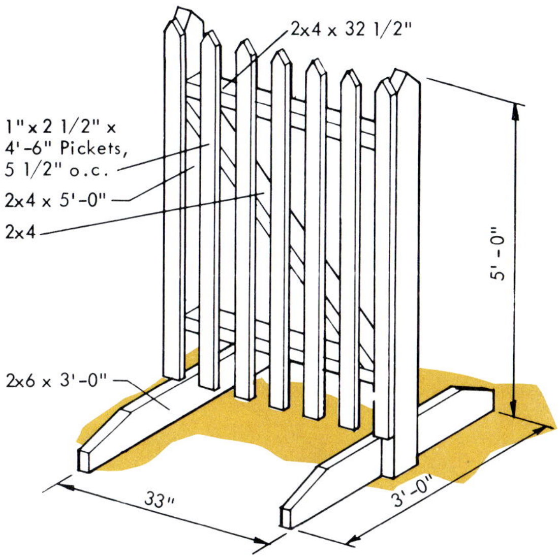

Wing

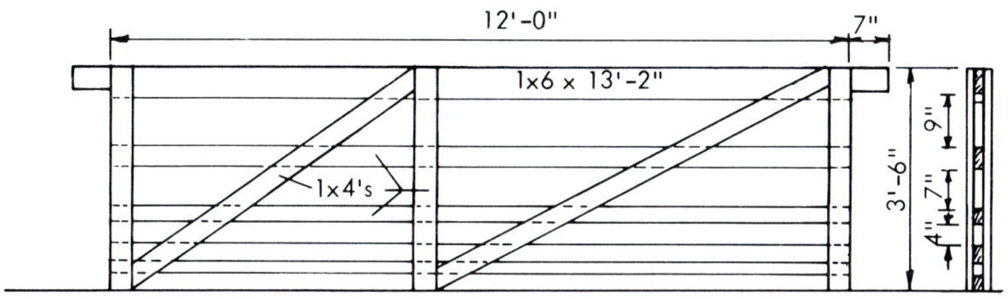

Gate

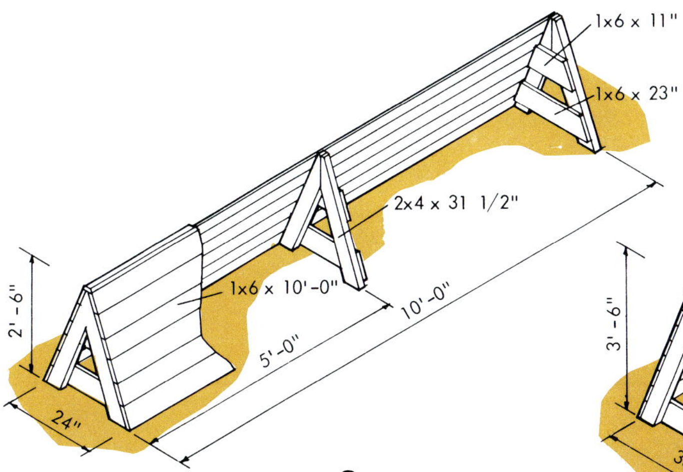

Coop

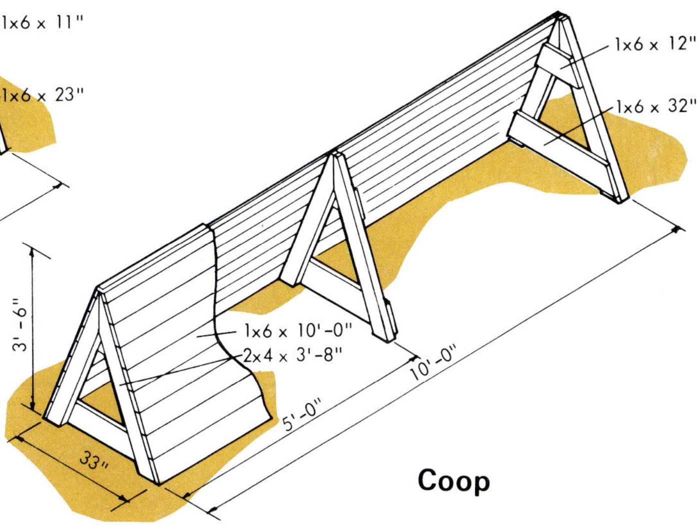

Coop

DUTCH DOORS

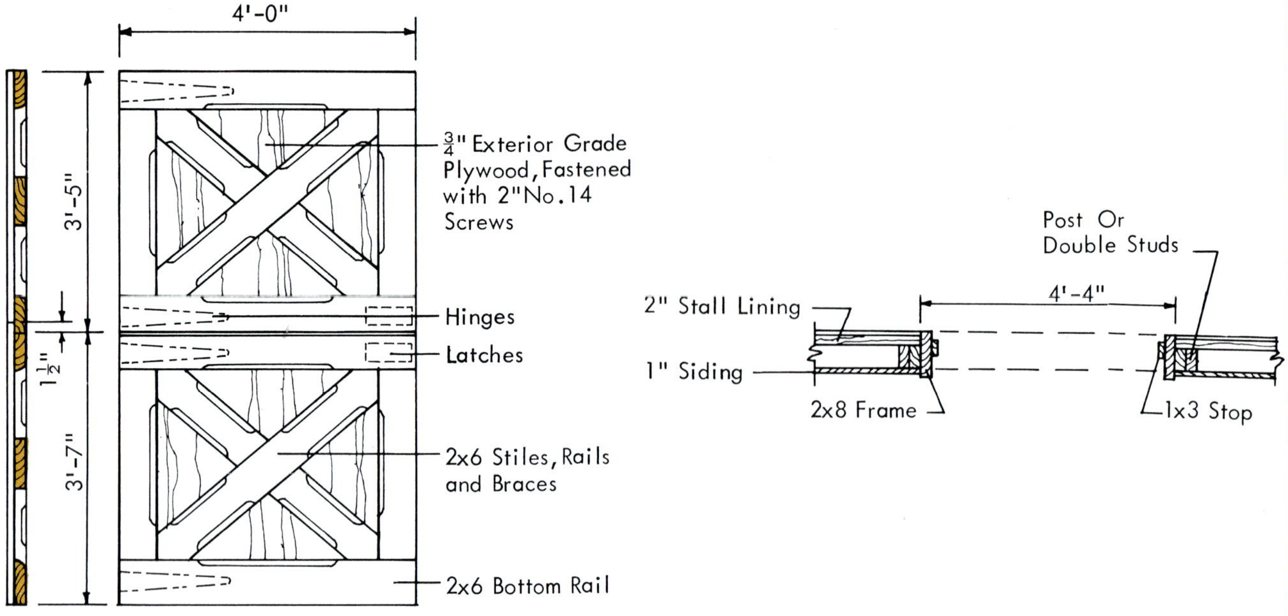

MINERAL FEEDER

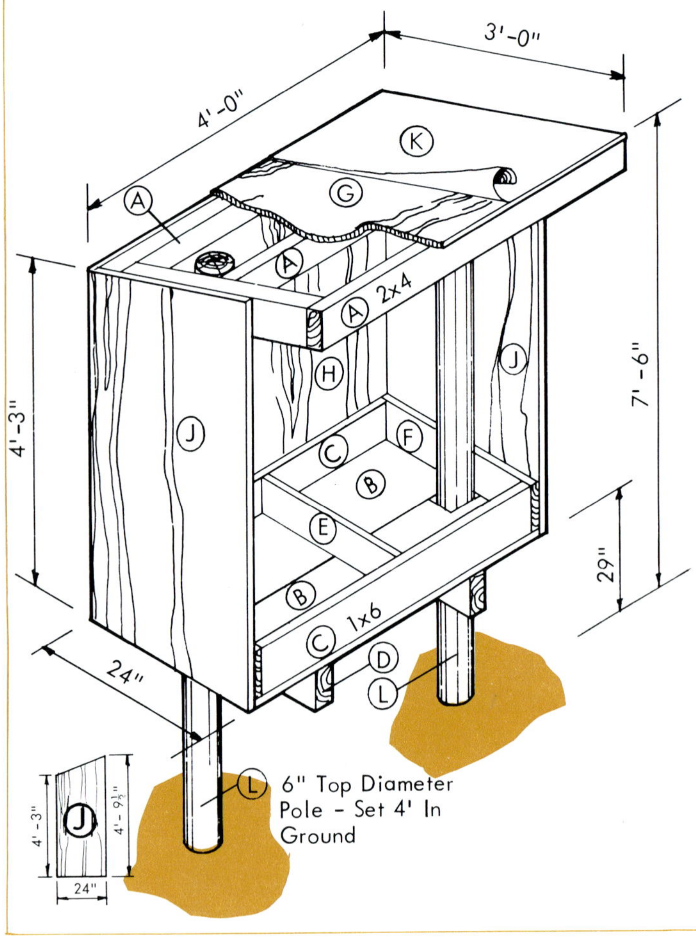

Cutting List

Item	No.	Description
A	2	2x4x4'-0"
B	2	1x12x3'-10½"
C	2	1x4x3'-9"
D	2	2x4x23"
E	1	1x6x21½"
F	2	1x6x23"
G	1	½"x3'-0"x4'-0" Ext Plywood
H	1	½"x4'-0"x4'-3" Ext Plywood
I	1	2x4x3'-9"
J	2	See Detail
K	1	3'x4' Roll Roofing
L	2	6" Dia Post, 11' Long

CUPOLAS

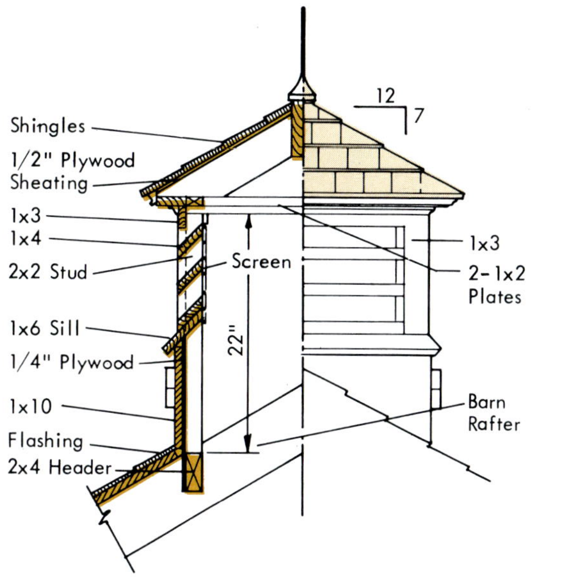
Shingles
1/2" Plywood Sheating
1x3
1x4
2x2 Stud
1x6 Sill
1/4" Plywood
1x10
Flashing
2x4 Header
Screen
22"
12
7
1x3
2-1x2 Plates
Barn Rafter

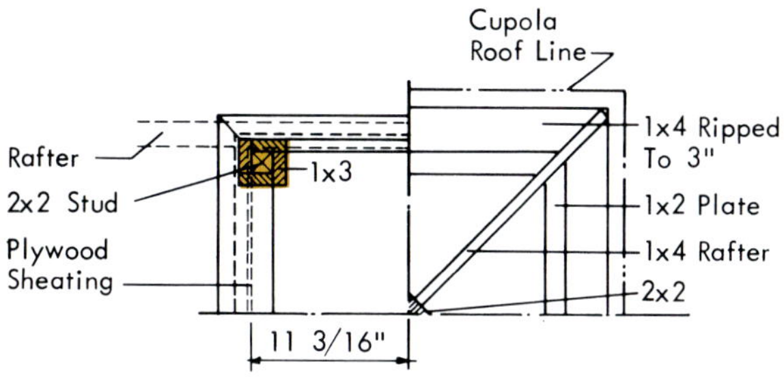
Cupola Roof Line
Rafter
2x2 Stud
Plywood Sheating
1x3
1x4 Ripped To 3"
1x2 Plate
1x4 Rafter
2x2
11 3/16"

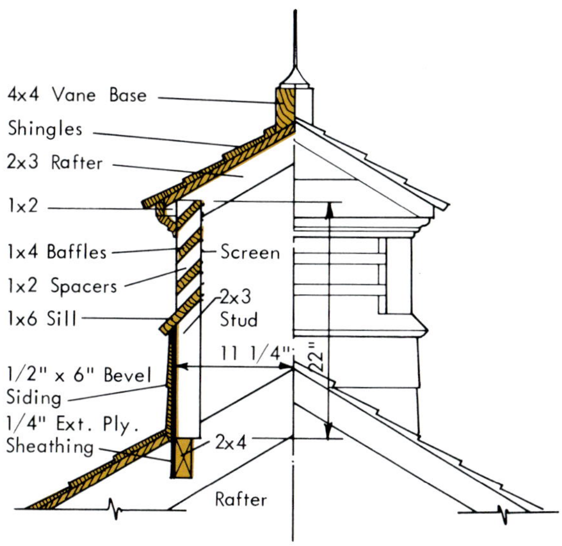
4x4 Vane Base
Shingles
2x3 Rafter
1x2
1x4 Baffles
1x2 Spacers
1x6 Sill
1/2" x 6" Bevel Siding
1/4" Ext. Ply. Sheathing
Screen
2x3 Stud
11 1/4"
22"
2x4
Rafter

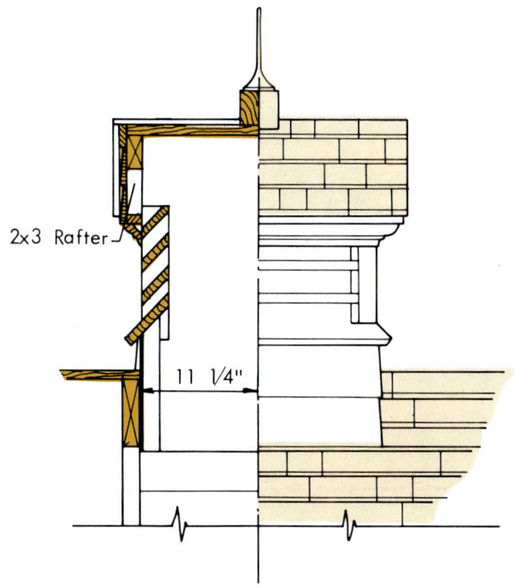
2x3 Rafter
11 1/4"

LOADING CHUTES

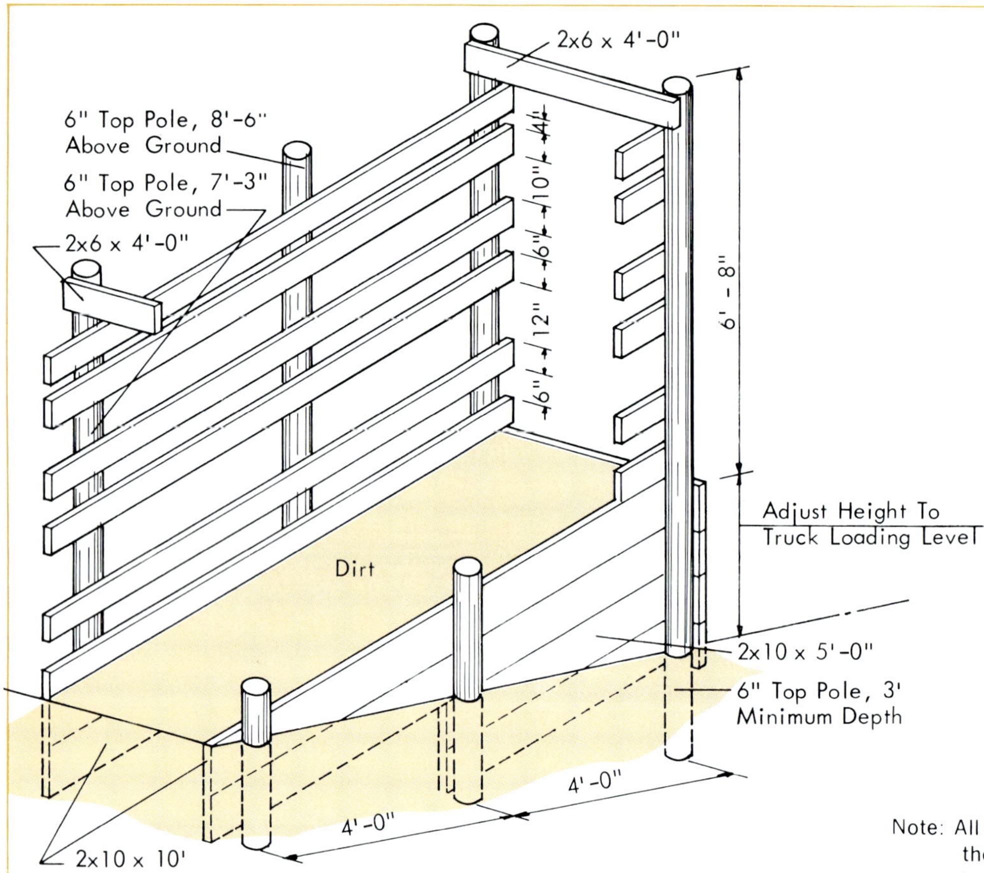

Bill Of Materials

No.	Description
2	6″ Top Pole, 14′ Long
2	6″ Top Pole, 12′ Long
2	6″ Top Pole, 11′ Long
4	2x10x10′-0″
4	2x10x5′-0″
4	2x10x4′-0″
2	2x6x4′-0″
12	2x6x10′-0″

Note: All poles and wood in contact with the ground should be pressure-treated

Note: Most horses are loaded from ground level or from loading dock at truck bed height.

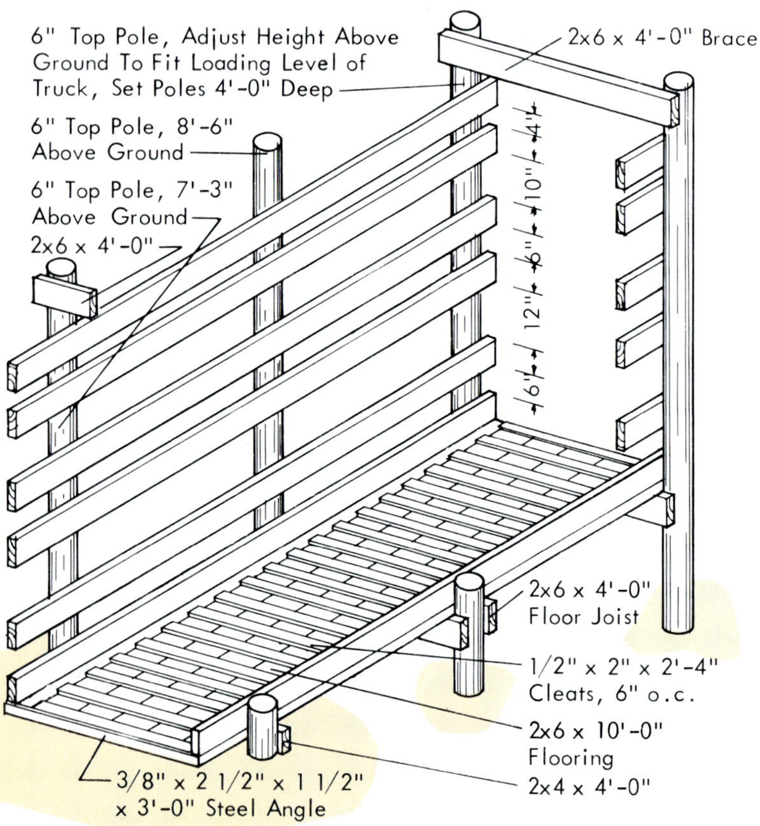

Bill Of Materials

No.	Description
2	6″ Top Pole, 14′ Long
2	6″ Top Pole, 12′ Long
2	6″ Top Pole, 11′ Long.
3	2x6 x 4′-0″ Joist
1	2x4 x 4′-0″ Joist
12	2x6x10′-0″
18	½″x2″x28″ Cleats
6	2x6 x 10′-0″ Flooring
1	3/8″ x 2½″x1½″x3′-0″ Steel Angle
2	2x6 x 4′-0″ Brace

SHOW BOX

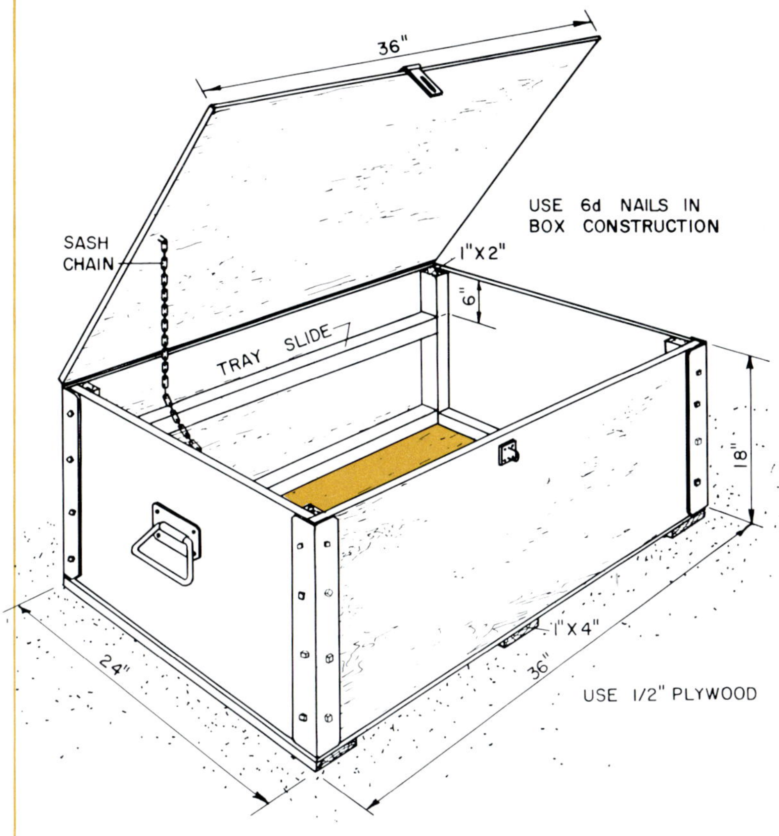

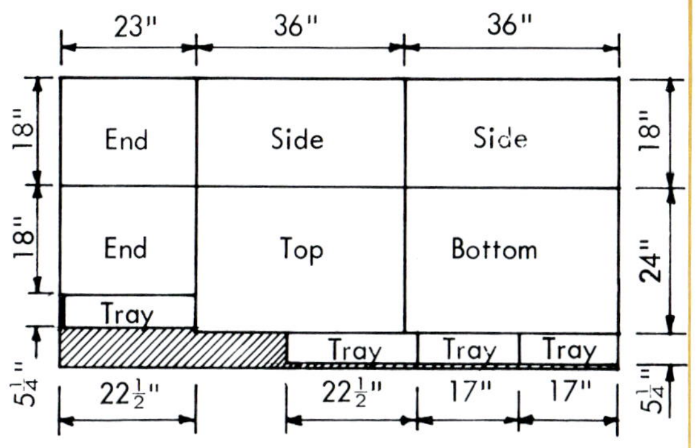

Cutting Diagram

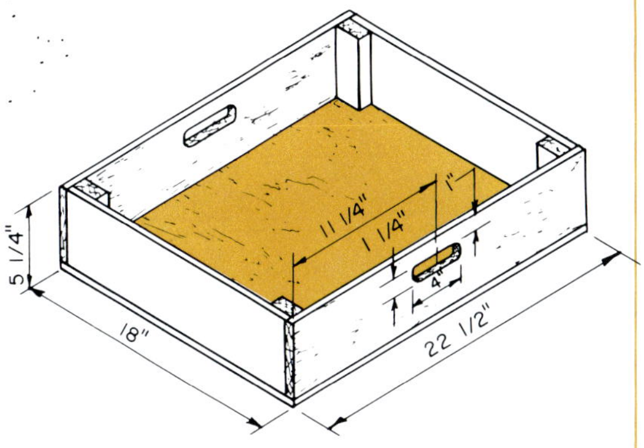

Tray

BREEDING STALL

A ¾'' rope is attached to Loop A, brought behind the mare above the hocks, and wrapped twice around the end of the pipe at the mare's right shoulder. The end of the rope is securely held by an attendant who can release it if necessary in an emergency.

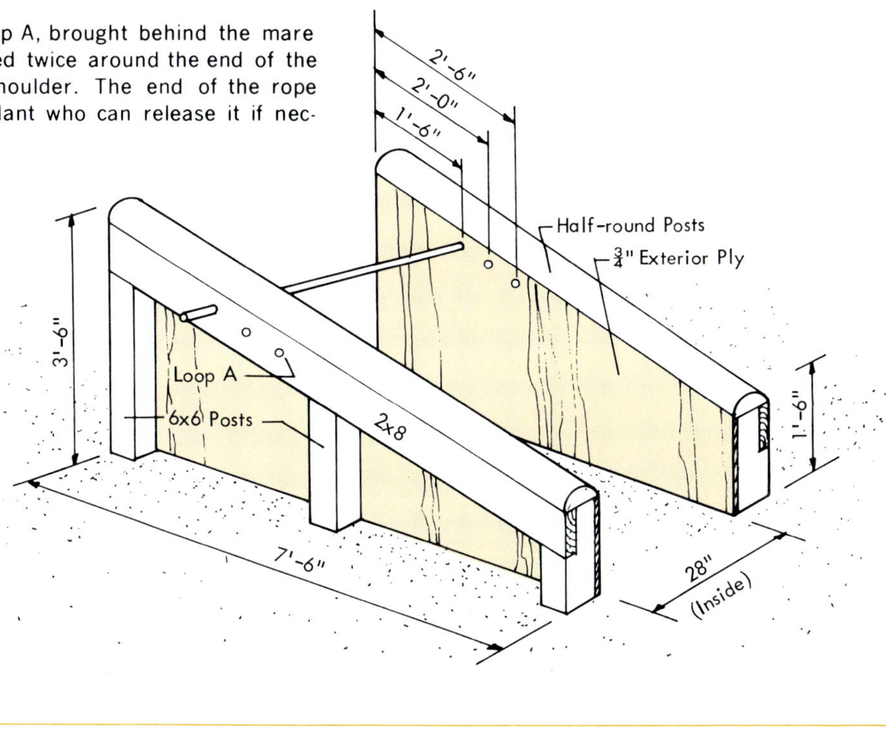

MATERIALS

Wood

Hardwoods come from broadleaf trees and are used primarily for fences and exposed interior surfaces subject to rough treatment.

Softwoods come from needle-leafed, cone-bearing trees and are used for structural framing and many non-structural uses.

Common boards are of nominal 1" thickness and are 4", 6", 8", 10", and 12" wide. Boards are available with a square edge, tongue and groove, or with a shiplap joint.

Square **Tongue and Groove** **Shiplap**

Structural lumber is nominally 2" thick and 4", 6", 8", 10", and 12" wide. Standard lengths are 4'-18' in multiples of 2'. Longer lengths can be purchased, but usually at a premium. The most common construction woods are pine, Douglas fir, hemlock, and spruce.

Rough lumber is full dimension and can give rustic beauty and added strength. It is usually more difficult to work with and finish than surfaced wood. Dimension lumber is surfaced smooth on four sides and is, therefore, less than full dimension.

Wood Preservatives

- Use commercially preservative-treated lumber.
- Buy and use lumber pressure-treated with oil-base preservatives for structural framing that will be exposed to soil, manure, foundations, or termite attack, and other framing pieces that would be expensive to replace.

The heartwood of all species has some natural decay resistance. Sapwood has low resistance and must be treated if long life is desired. The most resistant species are baldcypress, cedars, junipers, black locust, osage-orange, catalpa, and redwood. Even these species are preservative treated where high decay hazards exist, and where failure would require expensive repairs.

Common wood preservatives fall into three general classes: oils such as creosote, petroleum solutions of pentachlorophenol, and waterborne salts that are applied as water solutions. These preservatives are applied with or without pressure. More protection is gained with pressure than with brush, dip, or spray application. Oils are usually more protective than water-soluble salts.

Brushing or dipping oil preservatives is effective for wood that will be exposed to moisture for only short periods of time. All cutting, framing, and boring of holes should be done before treatment.

Brushing or dipping the wood with oil- or water-soluble solutions is seldom worthwhile for wood that will be exposed to weather, soil, or water.

Wood treated with water-soluble preservatives can be painted after the wood has dried. Wood treated with oils like creosote and penta can stain through paint unless the treated wood has weathered or been cleaned. If light oils are used for treatment, the wood can be painted as soon as the solvent oils have evaporated.

Do not use oil-preservative treated lumber where small children can come in contact with it. Protect hands and other parts of the body from direct contact with wet preservatives.

Hardboard

Hardboard is made of reconstituted natural wood fibers that have been pressed into panels. The panels have no grain, and one or both surfaces can be smooth. The surface is hard, and the installed panels are impact resistant if properly supported. The panels can be formed to curves. The panels are 4' wide and up to 16' long, and are 1/8", 3/16", 1/4", or 5/16" thick.

Hardboards may be tempered or standard. Tempered boards are stronger and have a higher moisture resistance than standard boards. Tempered boards are used for exterior and interior projects where moisture resistance is necessary. Standard boards are commonly used for interior walls and ceilings, and may be used for protected exterior surfaces. Follow manufacturers' application recommendations to avoid buckling due to moisture changes.

Plywood

Plywood for use outdoors, inside animal shelters, or where alternate wetting and drying may occur, should be exterior type.

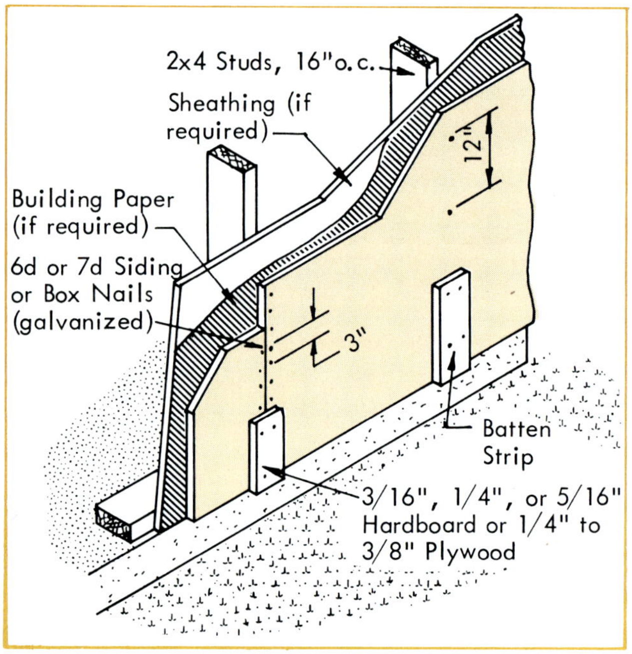

Using exterior plywood for exterior surfaces

Plywood is manufactured in two basic types, exterior and interior, and in a variety of appearance grades within each type. The glue bond and the grade of the plies determine the type. Exterior plywood has waterproof glue lines and all plies are at least grade C. Sheathing has waterproof glue lines and grade D exterior plies. Sheathing should not be used for outdoor uses unless covered with roofing or siding.

The most common plywood panel size is 4' x 8'. Standard thicknesses for sanded panels are ¼"-¾" in 1/8" gradations. Sheathing thicknesses range from 5/16"-3/4".

Sheet Metal

Most steel products will rust. Rusting causes the metal to slowly deteriorate and lose its strength. Rusting can be delayed or prevented with paint, zinc coatings (galvanizing), or by adding certain ingredients when making the metal.

Zinc coating of steel is common for products that will be exposed to moisture and weather. The thickness of the coating determines the length of time the product will withstand rusting. A 1¼-ounce coating is the standard weight for roofing and siding. For extra long life, *Seal of Quality* sheets with a 2-ounce coating are used. Galvanized metal can also be painted, which will add more life.

Iron-free metals do not rust, but will wear away because of weathering or corrosion. Weathering and corrosion can be delayed or prevented by covering with paint or other coating.

Steel and aluminum sheets are made in several shapes and thicknesses. Sheets are 26", 30", 36", or 48" wide. Common lengths are 5'-12'. The thickness and shape determine the strength of the sheet. Check with an engineer or follow an engineered plan before buying and using metal sheets.

Baked on colored finishes are available in metal siding. High quality material gives good service and makes an attractive building.

Attach sheets with nails or screws that seal the hole they make as they pass through the sheet. If you miss a framing member with a nail, pull the nail and fill the hole with a sheet metal screw. Use galvanized nails in steel (aluminum in aluminum) siding and roofing. The needed size and length of nail or screw depends on the shape of the metal sheet, the framing, and the strength the fastener must have. Follow the sheet manufacturers' recommendations.

Masonry Construction

Choose a material that will complement the surroundings. After choosing the building material (block, brick, or stone), choose a pattern that blends well with the site.

Make sure the dimensions will allow the use of full and half-length blocks.

Concrete masonry units are made in local plants and are available in most parts of the country. They are made in a number of sizes and shapes. Masons familiar with this type of construction are also available.

Before starting work consult the local products manufacturer. He will suggest how units can be used economically, and estimate the number and kind of each needed. Some plants can furnish lightweight concrete masonry units made with cinders, expanded slag, or burned shale.

The mortar mixes shown in Table 9 can be used in repairing and tuckpointing old masonry walls as well as for new masonry construction. After the mortar has partially stiffened, thoroughly compact the joints by tooling.

Use mortar within 2½ hours after original mixing when the air temperature is 80° or higher, and within 3½ hours when the air temperature is below 80°. Discard mortar not used within these time limits.

Table 9. Recommended mortar mix proportions by volume

Type Service	Cement	Hydrated lime	Mortar sand in damp, loose condition
For ordinary service	1–masonry cement*		2¼ to 3
	or		
	1–portland cement	½ to 1¼	4½ to 6
Subject to extremely heavy loads violent winds, earthquakes or frost action.	1–masonry cement* plus 1–portland cement		4½ to 6
	or		
	1–portland cement	0 to ¼	2¼ to 3

*ASTM Specification C91 Type II.

1. Build corners first, 4 or 5 courses higher than center of wall.
2. After laying each corner course, check alignment for level and plumb.
3. Use a 1x2 board with markings 8" apart to locate top of masonry for each course.
4. Use a mason's line stretched from corner to corner to insure horizontal accuracy.
5. Bring block to proper grade and make plumb by tapping with a trowel handle.
6. Run a round "O" or "V" shaped tool along joints after mortar has somewhat stiffened to help provide watertight construction.

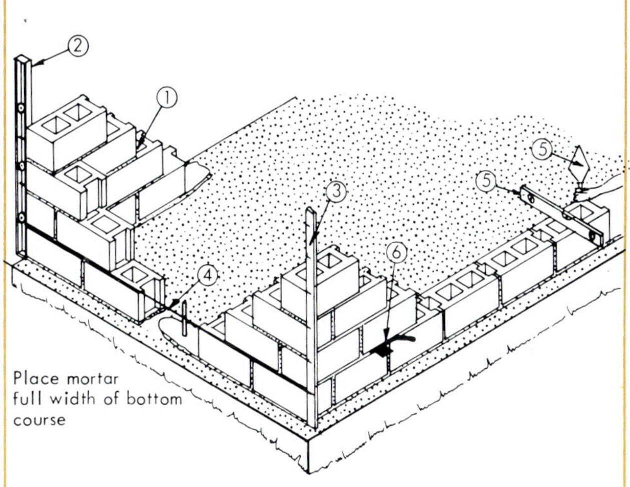

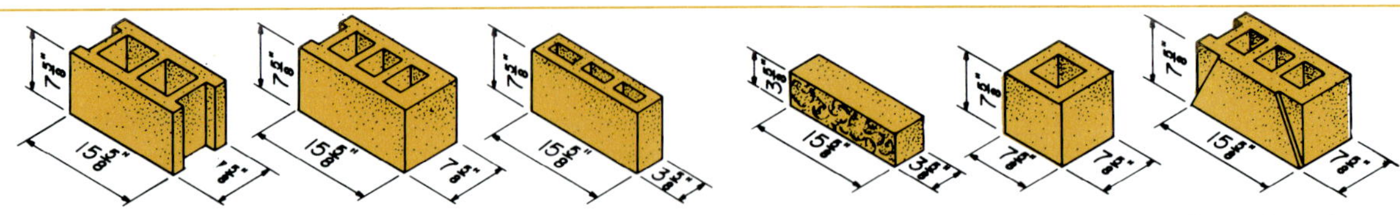

Common concrete block

Concrete

Concrete can be cast in almost any form. It is widely used, due to its fairly low cost, and its ease of handling in small batches. Concrete can be finished in either a smooth or rough texture. It can be painted or colored with pigments.

Concrete is a mixture of portland cement, water, and aggregates. Portland cement is sold in bulk, or in bags of one cubic foot (94 lb). The aggregates provide volume at low cost, comprising 66% to 78% of the concrete. See Table 10.

The cement and water form a paste which hardens and glues the aggregates together. The quality of concrete is directly related to the binding qualities of this cement paste.

Concrete is a durable material. However, if not properly proportioned, mixed, and installed, it will crack. It sometimes buckles when placed on a poor foundation.

Construction

Remove all sod and organic matter from the site. The sub-grade must provide uniform support and be easily drained. The top 6" of sub-grade should be sand, gravel, or crushed stone, where subgrades may be water soaked much of the time.

Use reinforcing wire in pavements poured over soft or spongy soils. Use 6x6 #10 wire mesh. Place the wire near the top of slabs that support light loads but are subject to frost heaving, and near the bottom of other slabs.

Before the concrete is placed, the subgrade should be thoroughly dampened. Place the concrete where it will be used to help prevent the scaling and dusting that results from too much handling. Spade or vibrate along the forms to eliminate voids or honeycombs.

Spread the concrete as soon as possible after it is mixed. After it has been dumped and spread,

Table 10. Concrete mixes

		[1]Gallons of water for each sack of cement, using:			[2]Suggested mixture for 1-sack trial batches			READY-MIX
						Aggregates[7]		
	Max. size aggregate	Damp[3] Sand	Wet[4] (average) Sand	Very[5] Wet Sand	Cement,[6] sacks (cu ft)	Fine, cu ft	Coarse cu ft	Sacks Cement Per Yard[8]
5-Gallon Mix; use for concrete subjected to severe wear, weather, or weak acid and alkali solutions.	¾"	4½	4	3½	1	2	2¼	7¾
6-Gallon Mix; use for floors (home, barn), driveways, walks, septic tanks, storage tanks, structural concrete.	1"	5½	5	4½	1	2¼	3	6¼
	1½"	5½	5	4½	1	2½	3½	6
7-Gallon Mix; use for foundation walls, footings, mass concrete, etc.	1½"	6¼	5½	4¾	1	3	4	5

[1]Increasing the proportion of water to cement reduces the strength and durability of concrete. Adjust the proportions of trial batches without changing the water-cement ratio. Reduce gravel to improve smoothness; reduce both sand and gravel to reduce stiffness.

[2]Proportions will vary slightly depending on gradation of aggregates.

[3]Damp sand will fall apart after being squeezed in the palm of the hand.

[4]Wet sand will ball in the hand when squeezed, but leaves no moisture on the palm.

[5]Very wet sand has been recently rained on or pumped.

[6]Use air-entrained portland cement in concrete for outdoor use. It virtually eliminates scaling due to freezing, thawing, and salt action. It is more workable and cohesive, reduces segregation and bleeding, and has improved sulfate resistance. Since it is more workable, it requires less mixing water—an added benefit.

Maximum Aggregate Size	Amount of Air
1½" - 2½"	4%-6%
1"	6%
3/4"	7%
1/2"	8%
3/8" or less	9%

[7]Aggregates must be clean and free of dirt, clay, coal, organic matter, etc. They must be durable, hard, and have few long slivers.

Fine aggregate, usually sand, should have particles up to ¼" in size. Coarse aggregates, commonly gravel or crushed rock, are pieces above ¼".

[8]Medium consistency (3" slump). Order air-entrained concrete for outdoor use.

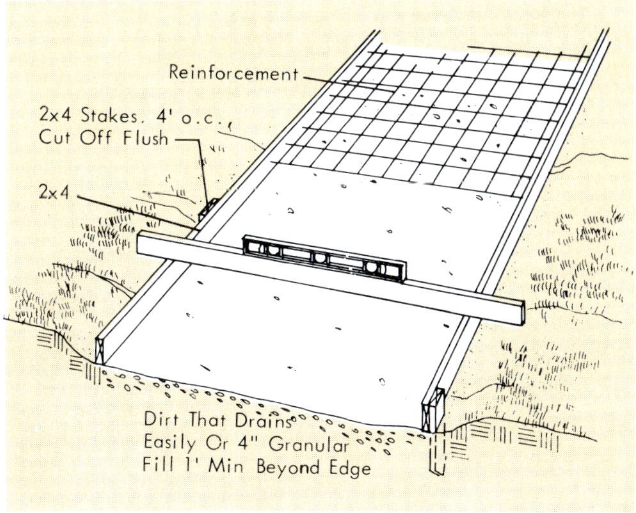

Preparing subgrades and laying reinforcing if needed

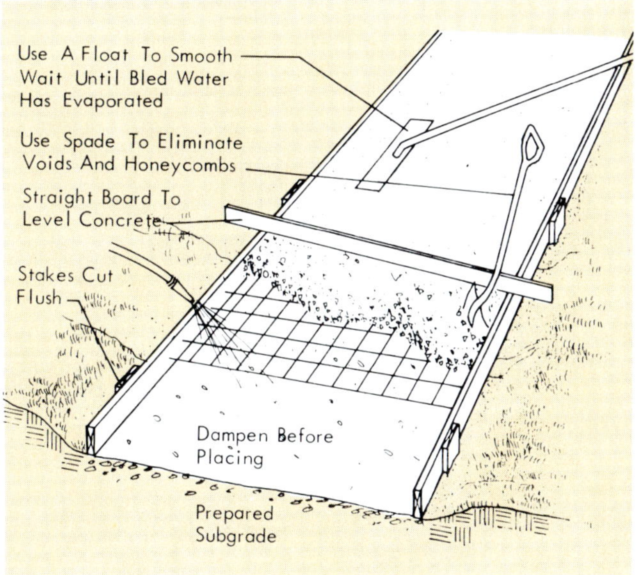

Placing the concrete

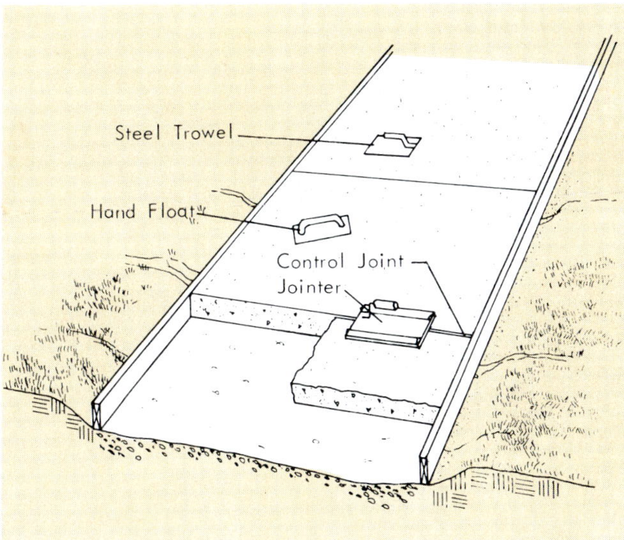

Control joints usually spaced at intervals equal to width of slab, but not more than 20 feet apart

strike it off to the proper grade, and then float to smooth and level the surface.

It is very important that placing, straightedging, and floating be completed before any bleeding occurs. (Bleeding refers to excess water in the concrete rising to the surface).

After all bleeding water has evaporated and the concrete has started to stiffen, start the other finishing operations. These include edging—rounding the edges to prevent chipping; and jointing—cutting a partial joint, called a control joint.

After edging and hand jointing, the slab should be floated. Floating will embed large aggregates just beneath the surface, remove slight imperfections and tool marks, and prepare the surface for other finishing.

The final smoothing is done with a steel trowel immediately following floating. If a rough surface is desired, steel troweling is omitted.

Control joints are cut across each slab to control cracking. Cracks, if they occur, will usually be in the joints. Cut control joints soon after the concrete has been placed to work coarse aggregates away from the joints. Space them about as far apart as the concrete strip is wide.

Isolation joints permit the slab to move with the earth. Place ¾" wide isolation joints along existing improvements such as buildings, concrete water tanks, or paved drives.

Expansion joints are constructed the same as isolation joints and should be installed in new walks and long drives.

Curing

Concrete does not dry—the paste sets by a chemical reaction between cement and water. Keep the surface of the concrete damp at least 5 days. Curing will continue for months. Remove forms after about 5 days for slabs, 10 days for walls, 28 days for structural elements.

Outside concrete work in cold weather

Concrete cures very slowly at temperatures below 50°. Water frozen in uncured concrete will expand and cause damage to the concrete. Concrete should be cured a minimum of 48 hours before it is permitted to freeze, but it is best to prevent freezing for 4 to 5 days.

Do not place concrete over frozen ground. Use Type III portland cement, or Type I with calcium chloride dissolved in the mixing water at the rate of 2 lb/bag of cement.

When air temperatures are below 40°:

- Heat the sand, gravel, and water to just below 150°. Heat the sand and gravel in separate piles over old culvert pipe, a section of smokestack, or other improvised firebox. Place a fire inside.
- Stir and rake the materials frequently to assure even heating.
- Always remove snow and ice from forms before placing concrete.
- Place the concrete right after mixing when the concrete temperature should be 60°-80°.

- Protect the placed concrete with a cover to retain as much heat as possible. Canvas, straw, or hay are often used as covers. Protect for 4 or 5 days.
- Maintain concrete temperature at 70° for 3 days, or 50° for 5 days. Do not allow it to freeze during the next 4 days. Use a vented heater indoors.
- Remove forms only after sufficient curing. Pour hot water on the concrete. If properly cured, there will be no effect, but if frozen, the concrete will soften.

Outside concrete work in hot weather

As the temperature rises above 70°, the curing rate increases. Evaporation of water from the concrete also increases. A combination of wind, high temperature, and low humidity will dry concrete too rapidly and weaken it. Keep the fresh concrete damp for at least 5 days, and place a sheet of 4 mil plastic over the dampened concrete to help retard evaporation.

In extremely hot weather it may be necessary to reduce the temperature of the freshly mixed concrete. Aggregates should be stockpiled in the shade, if possible, and cool water used to mix the concrete. Curing must be started promptly to retard the evaporation of water.

Paint

There are three basic steps to painting. The first is to select the right paints for the various surfaces to be painted. The surfaces must then be adequately prepared, and finally, the selected paint must be properly applied.

Selecting paints

Choose paints that have the desired color, can be applied successfully over the material needing paint, and can withstand the weathering or special conditions in your locality. Most local paint dealers know of any special conditions. It is good economy to buy a quality paint to lengthen the time between repaintings, and reduce costs.

Wood and most metals can be painted with either solvent- or water-base paint. But first, these materials require a primer coat to seal the surface against moisture.

Rough-sawed lumber, dense smooth woods, knotty or flat grain boards, or plywood of oak, pine, spruce, or fir can sometimes be more economically stained with a heavy pigmented oil stain than painted.

Use only lead-free paint on equipment to be used where children or animals may lick or eat the paint.

Some lumber used for siding is primed at the factory and does not need a prime coat. Follow the lumber manufacturer's recommendation on the paint to use. In most cases, they recommend a standard house paint.

Chalk-resistant paints should be used where chalk rundown would discolor masonry or metal. Where rundown would not be objectionable, the self-cleaning chalking paints can be used.

Porches, steps, and screens should be painted with specially formulated abrasion-resistant paints.

Special solvent-thinned trim paints are available for wood trim, screen frames, shutters, and other small areas of a building.

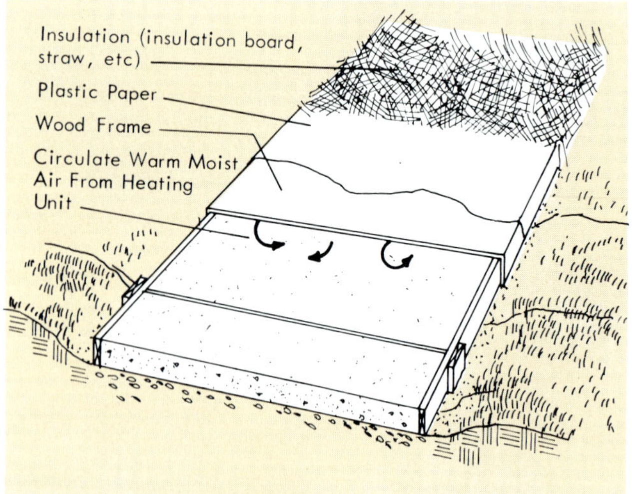

Protecting concrete in cold weather

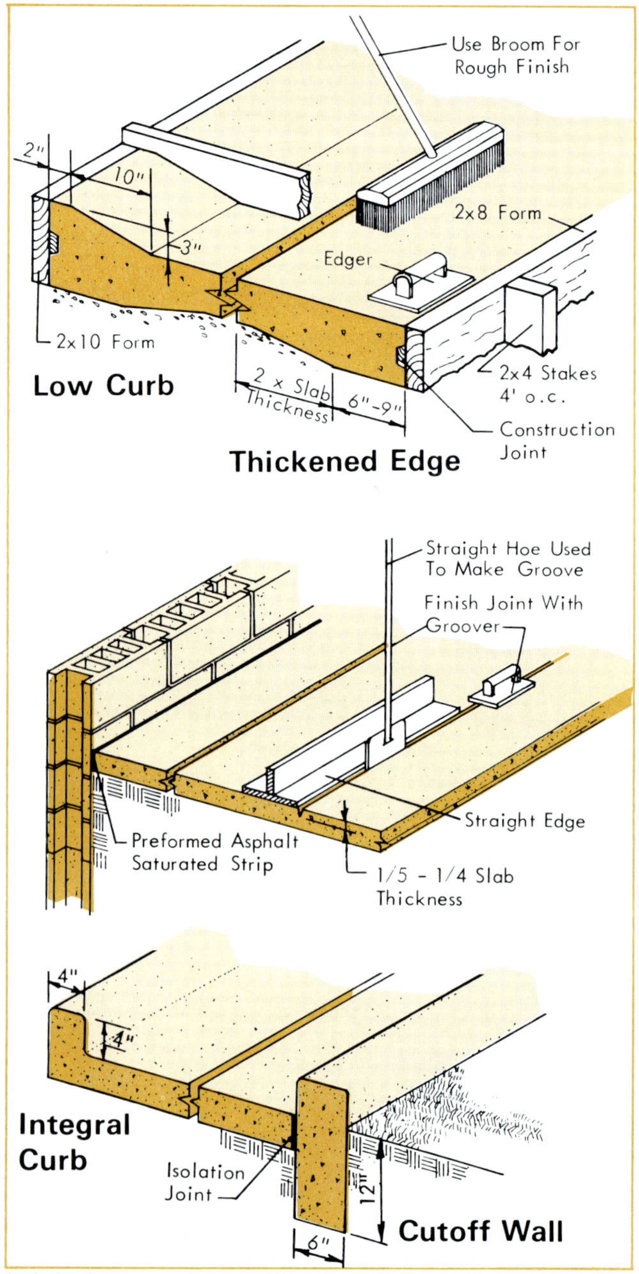

Isolation joints permit the slab to move with the earth. Place isolation joints along existing improvements such as buildings or paved drives

There are two types of natural finishes—surface coating and penetrating. Surface coatings are not very durable and must be refinished every year or two to prevent discoloration of the wood. Some penetrating finishes are very durable.

Preparing to paint

Wood surfaces

All surfaces must be cleaned of dirt, oil, scale, and other impurities. Surfaces should be cleaned and dusted just prior to painting.

Metal surfaces

Remove all rust, oil, and grease, and smooth roughened surfaces. A wire brush, steel wool, or sandpaper will usually remove rust and rough areas. Remove oil and grease with solvents such as mineral spirits, and then rinse thoroughly. A primer must be applied immediately after cleaning.

Masonry surfaces

Remove all dirt, loose particles, oil, efflorescence, and old paint that is peeling or heavily chalked. Efflorescence is a white salt-like material adhering to concrete or mortar and is caused by salts within masonry coming to the surface. Dirt, efflorescence, and loose particles can usually be removed by washing the surface with a cleaner or detergent and water.

Poured or precast concrete may have a "release agent" on the surface which must be removed with a solvent or by several months of weathering.

Masonry floors, decks, porches, and patios may require a special treatment to provide an adequate base over which to apply paint. Manufacturers of paint for this kind of surface usually give instructions on how to prepare these surfaces.

Concrete floors should not be painted until the concrete has aged at least three months. Slick concrete floors should be roughened before being painted with one gallon of muriatic acid mixed in two gallons of water. Brush the acid into the floor until it fizzes in every area. Rinse the acid off with clear water, examine the floor, and retreat spots that were not roughened. Floors with an even coarse texture can be painted without an acid treatment.

Paint will not fill the pores of brick, block, or stucco. These surfaces must first be plastered if a smooth surface is desired.

Iron, copper, or other metallic objects embedded in masonry surfaces to be painted with water-thinned paints must be primed with a good anti-corrosive primer.

After all impurities are removed or primed, wash or hose the surface. Do not wash or hose the surface if efflorescence was present. Allow the surface to dry thoroughly if solvent-thinned paints will be applied. The surface need not be completely dry before applying water-thinned paints.

Application procedures

Apply paint in reasonably clear, dry weather with the air temperature above 40°, preferably 50° or higher. The paint temperature should be about 70° for easy stirring and spreading. Apply solvent paints only when the surface is dry. Several days' drying time may be needed after a rain. Water-thinned paints may be applied to a surface that is damp but free of liquid water. Do not paint during windy or dusty weather, or when insects may get caught in the paint coat.

In hot weather, paint surfaces after they have been exposed to the sun and are in the shade. If painting must be done in the hot sun, avoid over-thick application of the paint—brush the paint out well.

Allow a solvent-thinned prime coat to dry about three days in fair weather and four days in humid weather, unless the instructions on the label indicate a shorter time is satisfactory.

Apply the finish coats of paint immediately after the primer has dried sufficiently and all caulking has been done. If you cannot apply the finish coats within a month, wash the surface thoroughly before applying the top coats.

Allow about 48 hours drying time before applying a second solvent-thinned finish coat. Two coats of latex paints may be applied in one day.

Three coats of paint are recommended for new work—one primer and two finish coats. However, properly designed and properly applied two-coat systems will usually give good durability.

Painting wood

Most paint manufacturers recommend applying a prime coat of paint to seal the surface against moisture. Use a primer recommended by the paint manufacturer. Apply a prime coat to all new surfaces, to all repaired surfaces, to surfaces that have little old paint left, and to old surfaces that are very chalky. Prime as soon as possible after all the repairing and cleaning is completed, but before filling with putty.

The Forest Products Laboratory recommends applying a water-repellent preservative before priming new wood that has not been so treated. The preservative contains penta and should be allowed to dry for two warm, sunny days before applying the primer. Ask your paint dealer about the recommendation of his paint manufacturer.

Use a zinc-free oil-base primer for new wood and spot priming when the top coats will be a solvent-thinned paint. Most manufacturers recommend using a solvent-thinned primer when the top coat will be an exterior emulsion or latex paint.

After a surface is primed, apply any caulking that is needed. Examples: around window and door trim; where different materials come together—joints between flashing and bricks; joints in wood siding; and at the corners of buildings.

Painting metal

Paint only when the surface is clean and dry. Apply a primer to new metal and to metal that has had old paint removed. Good primers will have zinc-dust, red lead, zinc yellow, blue lead, oxide of iron, or zinc-dust oxide as one of its ingredients. Zinc-dust paints must be stirred frequently.

After the primer has dried sufficiently, apply one or two coats of finish paint. Most ordinary house or trim paints can be applied as finish coats over any of the recommended prime coats.

Painting galvanized metal

Follow the procedures above for preparing and painting metal. Zinc-dust paints in an oil, alkyd, or phenolic base adhere best to galvanized surfaces under most conditions. Cement paints in an oil base work nearly as well, and are recommended where color is desired. Latex paints made for galvanized steel are being used but have not been tested as extensively as other metal paints. These latex paints seem to be especially suitable for single coat coverage.

Forest Products Laboratory Stain

Ingredients and methods of application for a stained finish on wood are explained in detail in the Forest Products Laboratory publication, FPL-046. You can obtain a copy from the Forest Products Laboratory, USDA, Madison, Wisconsin 53705.

The stain, based on durable red and brown iron oxide pigments, adequately applied to rough surfaces, will last up to ten years. When refinishing is necessary, the rough surface-stain system is easily renewed without extensive surface preparation.

FASTENERS

Nails

Nails that have special coatings, such as zinc, are intended primarily for use where corrosion and staining may occur. Cement coated nails will increase the strength of a joint for a short time, but the joint strength will drop to the strength of a plain-nail joint in a few months.

Table 11. Selection of nails for use with different materials

	NAIL TO USE[1]
1″ stock	8d
2″ stock	16d to 20d
3″ stock	40d to 60d
Concrete forms	common or double headed nails
Toenailing studs	10d
Sheathing; roof, wall, and floor	8d
Roofing	
Aluminum	1¾″ to 2½″ aluminum nail with rubber washers.
Asphalt shingles	Large head roofing nail
Wood shingles	3d to 4d
Nailing steel sheet metal (roofing and siding)	Self-tapping screws, helical drive screws with lead washers.
Nailing to concrete	Concrete or cement nails or helical drive nails or drive bolts.
Plywood[2]	
3/4″ plywood	6d casing or 6d finishing
5/8″ plywood	6d or 8d finishing nails
1/2″ plywood	4d or 6d
3/8″ plywood	3d or 4d
1/4″ plywood	3/4″ or 1″ brads; 3d nails

[1] Recommendation: Use galvanized hardened threaded nails for most wood outdoor projects. Use aluminum nails with aluminum sheets.

[2] The nailing strength of plywood is about the same as solid wood, but the greater resistance to splitting when nailed near the edge is a definite advantage.

Table 12. Nails commonly used and their approximate strength in pounds

SIZE	LENGTH INCHES	WIRE GAGE	APPROX. NO./LB.	APPROX. STRENGTH POUNDS PULL (1)	LATERAL (2)
COMMON NAILS					
2d	1	15	847	Douglas Fir, Larch	
3d	1 1/4	14	543	or Southern Pine	
4d	1 1/2	12 1/2	294		
5d	1 3/4	12 1/2	254		
6d	2	11 1/2	167	29	63
7d	2 1/4	11 1/2	150		
8d	2 1/2	10 1/4	101	34	78
9d	2 3/4	10 1/4	92		
10d	3	9	69	38	94
12d	3 1/4	9	63	38	94
16d	3 1/2	8	49	42	107
20d	4	6	31	49	139
30d	4 1/2	5	24	53	154
40d	5	4	18	58	176
50d	5 1/2	3	14	63	202
60d	6	2	11	68	223
SPIKES					
10d	3	6	32	49	139
12d	3 1/4	6	31	49	139
16d	3 1/2	5	24	53	155
20d	4	4	19	58	176
30d	4 1/2	3	14	63	202
40d	5	2	12	68	223
50d	5 1/2	1	10	73	248
60d	6	1	9	73	248
5/16	7	5/16″	6	80	289
3/8	8-12	3/8″	5-3	96	380
HARDENED THREADED NAILS					
6d	2	12	190	80	69
8d	2 1/2	11	117	90	82
10d	3	10	78	100	94
12d	3 1/4	10	73	100	94
16d	3 1/2	9	57	110	107
20d	4	7	36	135	139
30d	4 1/2	7	31	135	139
40d	5	7	27	135	139
50d	5 1/2	7	23	135	139
60d	6	7	18	135	139

(1) Per inch penetration of point
(2) For penetration of 11 diameters

In general, annular-grooved and spiral-grooved nails will give stronger joints than plain nails. The strength of the nail is greatest when it is driven perpendicular to the grain in the wood.

Toenailing—slant driving nails through the edge or end of a board—is most effective when

- the longest nail that will not cause excessive splitting is used,
- the entry point is about one-third the length of the nail,
- the nail is driven at a 30 degree slope, and
- the shank of the nail is buried without excessive damage to the wood.

Nails will be strongest when driven into lead holes that are slightly smaller than the nail. Lead holes also prevent or reduce splitting of the wood and are used mainly in hardwoods.

Spacing

Space nails about 6" apart for most work. Nails with glue produce a strong, durable joint. The nails secure the joint until the glue cures.

Predrilling lead holes may be necessary for nails placed near edges of boards. The drill bit should be slightly smaller than the nail.

Wood Screws

Lubricating the surface of a screw with soap is recommended to facilitate insertion, especially in dense woods.

Screws should always be turned in. They should never be started or driven with a hammer as this will tear the wood fibers and injure the screw threads, seriously reducing the load carrying capacity of the screw.

Table 13. Screw dimensions

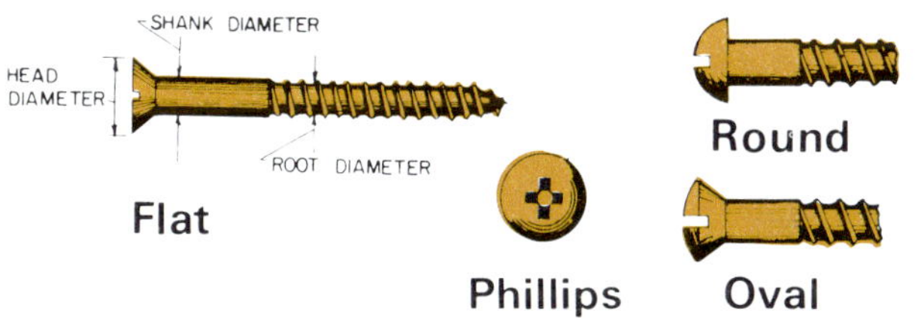

	NO. OF SCREW	SHANK HOLE SIZE	THREADS ROOT DIAM.	THREADS lead hole SOFT	THREADS lead hole HARD	THREADS PER INCH
	0	1/16	.040	1/64	1/32	32
	1	5/64	.046	1/32	1/32	28
	2	3/32	.054	1/32	3/64	26
	3	7/64	.065	3/64	1/16	24
1'' Lumber	4	7/64	.075	3/64	1/16	22
1'' Lumber	5	1/8	.085	1/16	5/64	20
1'' Lumber	6	9/64	.094	1/16	5/64	18
1'' Lumber	7	5/32	.102	1/16	3/32	16
1'' Lumber	8	11/64	.112	5/64	3/32	15
	9	3/16	.122	5/64	7/64	14
2'' Lumber	10	3/16	.130	3/32	7/64	13
2'' Lumber	11	13/64	.139	3/32	1/8	12
2'' Lumber	12	7/32	.148	7/64	1/8	11
2'' Lumber	14	1/4	.165	7/64	9/64	10
2'' Lumber	16	17/64	.184	9/64	5/32	9
2'' Lumber	18	19/64	.204	9/64	3/16	8
	20	21/64	.223	11/64	13/64	8
	24	3/8	.260	3/16	7/32	7

Which screw to use

For maxiumum strength, screw threads should penetrate seven shank diameters, or about seven times the recommended shank lead hole.

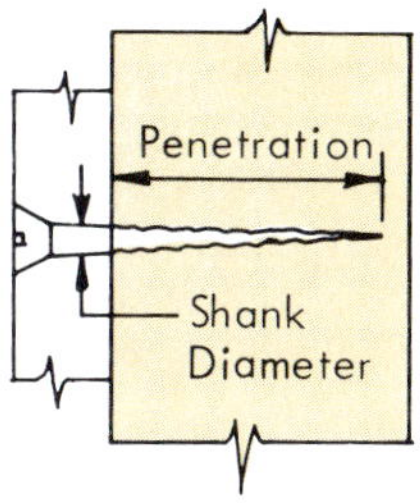

Bolts

Bolts are used when great strength is required or when the equipment must be disassembled frequently. A washer must usually be placed between the nut or head and a surface to prevent loosening, to allow for long life and strength of the joint, to avoid marring the surface, and to permit easier tightening.

Carriage bolts are usually for wood-to-wood applications, and are from 1/4″-1″ in diameter and from 3/4″-20″ long.

Machine bolts have National Fine or National Coarse threads, are from 1/4″-3″ in diameter, and come in almost any length. The hole through which the bolt is placed should be 1/32″-1/16″ larger than the bolt.

Stove bolts are made with either flat or round slotted heads and may be threaded over their full length. They are generally used with square nuts and are used for many applications. If flatheaded, they are counter-sunk; if roundheaded, they are drawn flush to the surface.

An *expansion bolt* is a bolt used with an expansion shield to provide anchorage in substances where a threaded fastener alone is useless. The shield is inserted in a predrilled hole, expands when the bolt is driven into it, and becomes wedged firmly in the hole.

Bolt types

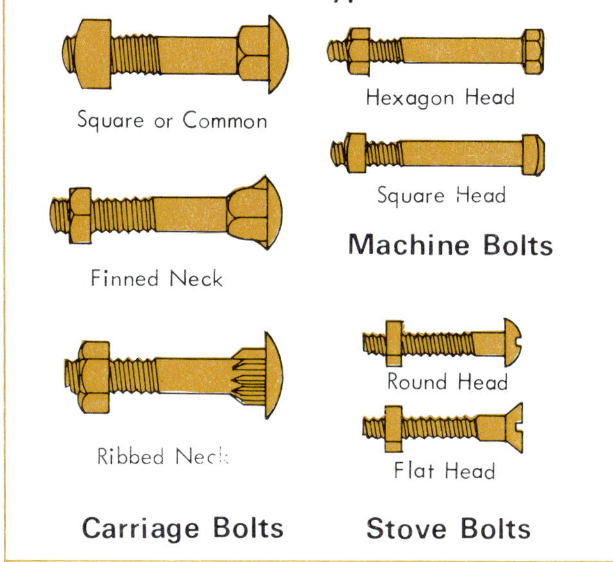

Glue

Gluing two pieces of wood is similar to welding two pieces of steel. The joint itself is stronger than the wood. If something breaks, it will usually be the wood, not the glue.

The wood to be glued should be dry, smooth, and free of dirt, oil, and other coatings. Most purchased lumber has been surfaced on a planer and is usually sufficiently smooth.

To clean dirt, paint, and other coatings from wood, use a scraper, wire brush, steel wool, or other abrasive. Do not use wood that has oil or grease at the joint location.

Generally, preservative-treated wood must be planed prior to gluing to obtain maximum holding power. Wood treated with oil-base preservatives tends to bleed. Buy wood that has been steamed or otherwise cleaned until bleeding has stopped.

Selecting a Glue

Pick the glue for the particular job and **Learn** how to use it.

To find a glue that will satisfy all requirements is impossible. Glue should be chosen because it will satisfy the requirements of one job. No glue is foolproof. Well-made joints require woodworking skill and close control of the gluing conditions. Improper use of glues and glued products may result in costly repairs or replacements.

Two glues are recommended for outdoor equipment: Casein and Resorcinol Resin.

Resorcinol Resin can be used for both wet or dry conditions. Apply this glue at 70° or above. Assemble, but wait 5 to 10 minutes before applying pressure. Maintain pressure for 10 to 16 hours.

Applying Glue

Before applying glue, be sure of a good fit by testing the joint; both pieces should make contact at all points. Apply glue with a brush or paint roller. Put pressure on the joint with clamps, nails, screws, or other fasteners before wiping off excess glue.

Pressure is usually applied with nails or staples. The commonly used nails are box, galvanized, or cement coated. The nails are not removed after the glue has cured. If staples are used, finish driving the staple with at least one hammer blow. At the end of the pressing time, the glued elements may be moved but should not be used for about one week.

Enough glue must be applied to the joint. When pressure is applied, some glue should ooze out from around the joint. If it doesn't, you aren't using enough glue. Never skimp on the use of glue; the cost of the glue is a minor item in the total cost of construction.

Joints connecting structural framing are as important as the framing itself. A pressure of one or more nails every eight square inches of the joint is usually required. The best location and size of pressure nails are usually given in a designed plan or in the engineer's recommendations.

Table 14. Properties of different glues

PROPERTY	SYNTHETIC RESIN GLUES					PROTEIN GLUES	
	Resorcinol	Urea	Polyvinyl	Epoxy	Contact Adhesives	Casein	Animal
Needs Mixing	X	X		X		X	X
Crack-Filling			X	X	X	X	
Applied Hot							X
Applied Cold	X	X	X	X	X	X	
Colorless Glue Line		X	X		X	X	X
Dark Colored Glue Line	X						
Tends to Stain Certain Woods						X	
Pressed at 70°	X	X	X	X	X	X	X
Over 8-Hour Working Life		X	X				X
Low Moisture Resistance			X				X
Medium Moisture Resistance			X		X	X	
Good to High Moisture Resistance	X	X		X			
Low Temperature Resistance		X	X				
High Temperature Resistance	X			X	X	X	X
For Structural Gluing	X					X	
For Exterior Uses[1]	X			3			
For Interior Uses[2]		X	X	3	4	X	X

1Exterior uses include outdoor furniture, boats, and recreational equipment.

2Interior uses include furniture, cabinets, framing, and other shopwork that will be used in a moderately dry atmosphere.

3Excellent for bonding metal, plastics, and cloth to wood. No practical advantage on wood-to-wood gluing over resorcinal resin except it is a good joint filler.

4Covering counters and cabinets with leather, linoleum, and plastic laminates.

BUILDING CONSTRUCTION

Use only quality materials and good construction practices. The initial cost may be somewhat greater, but good quality will increase the life of the building, defer maintenance and replacement costs, and also contribute to such intangibles as personal satisfaction and pride of ownership.

The kind of building shell used will depend on:
- The building's function.
- The size of the structure.
- Local, state, and national codes.
- Personal preferences.
- Available funds.

Design and Construction of the structure and its foundations must be adequate to support the weight of the complete building, and the forces applied to it. The loads that affect most buildings are:
- The weights of materials used in construction.
- Snow load on the roof.
- Wind loads on the roof and walls.
- Weights of materials stored in the building and supported by it.

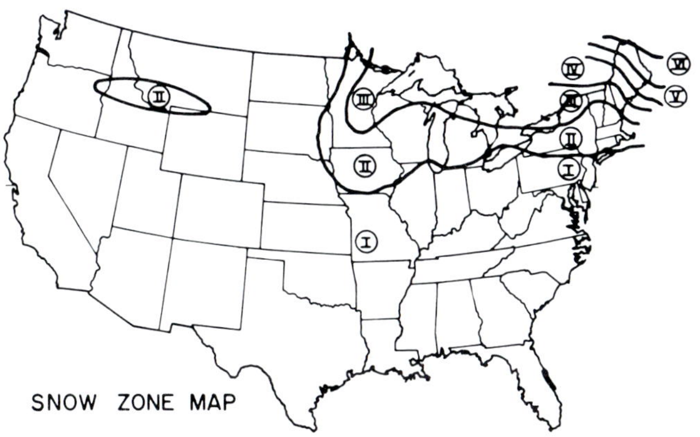

SNOW ZONE MAP

Snow loads and wind loads vary between states and are often different in various parts of the same state. While it is not practical to make buildings strong enough to resist tornadoes, they can be protected from severe damage from high winds by using preventative construction practices:
- Use foundations designed to resist overturning and lifting.
- Anchor the superstructure so it cannot be torn loose from the foundation.
- Make all joints between framing members strong enough to resist horizontal forces and uplift.
- Brace all walls to resist horizontal pressure and suction.
- If possible, locate buildings where they are protected from storms by hills or windbreaks.
- Use good materials and workmanship. Apply materials such as roofing in accordance with manufacturers' recommendations. Use mortar with not more than 3 cu ft of sand per sack of cement. Use pressure-treated wood for poles set in the ground and for other wood in contact with the ground or foundations.

The foundation must be designed for the building it will support and the site it will occupy. It must resist the forces acting on it and the action of the soil below and around it:
- Building weight and roof snow load.
- Wind uplift and overturn forces.
- Soil load bearing value at the building site.
- Soil movement caused by changes in soil moisture content.
- Heaving during freezing and thawing of the soil.

Construction Considerations

Most contractors are knowledgeable about building materials and their uses. Most builders like to erect good, attractive, buildings. But, the buyer must select the style, type, size, and quality of construction that fit his specific needs for the present and the forseeable future.

Decisions about facilities for horses are more than engineering inputs to business profits. Because most pleasure horses are owned for recreation, and are therefore non-profit personal enterprises, decisions are usually made to get the most satisfaction for the money spent in terms of personal taste and pride of ownership.

But, whether horses are owned for profit or pleasure, buildings and facilities should be in balance with the rest of the enterprise by contributing as much or more than can some other alternative. They can be out of balance because of:
- Underimprovement—performance and appearance could have been improved by investing additional effort and money,
- Overimprovement—needs could have been fulfilled with less investment, or
- Misplaced improvement—the buildings and facilities fail to perform as expected, regardless of quality of construction and appearance, because of under or over capacity, poor location, or because they do not fit the needs of the enterprise.

The most effective way to get a needed building is to select an engineered building plan with specifications that fit your situation, and obtain competitive bids from several reputable builders. Having a plan developed for you can be time consuming. Also, many builders and building manufacturers have their own standardized plans which they prefer to use, but which may not match your situation. Beware of plans that do not fit your need, and also of the builder who cannot supply a detailed plan of the proposed building. Most reputable custom builders prefer to bid on and work from an engineered plan and a good set of specifications.

CONSTRUCTION PLANS

The drawings in this book are more or less typical of the facilities associated with horse care and use. Except for a few equipment items, the illustrations should not be used as working drawings. They are intended to give information helpful in the preparation of working drawings. Working drawings should be prepared by a qualified agricultural engineer or architect to conform with local codes and regulations, and to adapt to local climatic and soil conditions.

The Midwest Plan Service does not have horse barn plans, although a number of their plans can be adapted for related facilities. However, most states distribute USDA plans for horse barns and equipment.

Current USDA plans are usually for uninsulated buildings for warm or moderate climates. With structural changes for heavier snow loads and the use of adequate insulation and ventilation, and in some instances supplemental heat, the USDA plans can be converted for cold climate use and for warm buildings.

Direct questions regarding the adaptability of the USDA plans for the north central region to the Extension Agricultural Engineer of your State Agricultural University—see inside the front cover.

Portable One-Horse Stable USDA 6082

This portable, one-horse stable is 10' x 12' and is planned for economical use of materials with a minimum of waste. The wall panels can be largely framed and finished on the ground, then tilted up and secured to the pressure-treated timber base. In areas subject to high winds, the stable should be securely anchored in place.

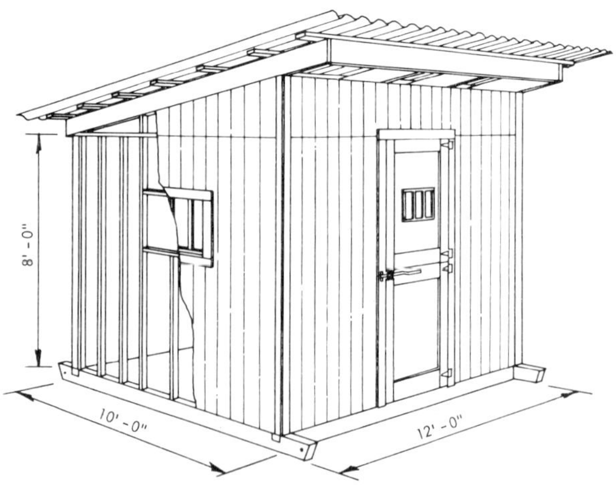

Two-Stall Barn USDA 5838

Two 12' x 12' box stalls with clay floors, a 6' x 8' tack room, and an 8' x 12' feed room are features of this 22' x 34' horse-barn. There is a useful covered way, and Dutch doors provide entry to the stalls. The barn may be expanded to house more horses.

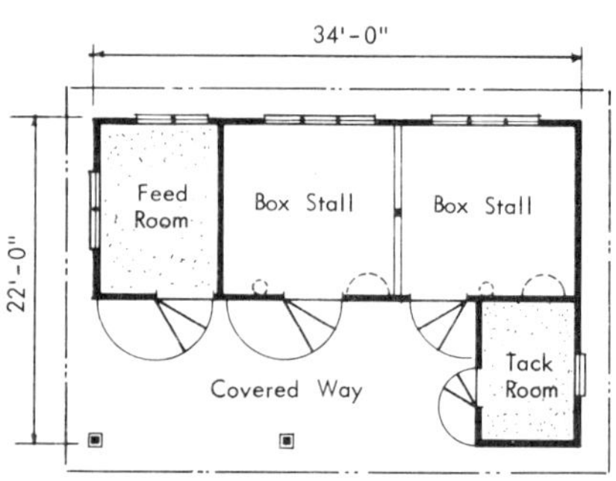

Three-Stall 1½ Story, Barn USDA 6024

This plan features three stalls, a feed room, a tack room and a 12' x 36' working alley. There is ample overhead storage for hay and bedding. The barn is 24' x 48'.

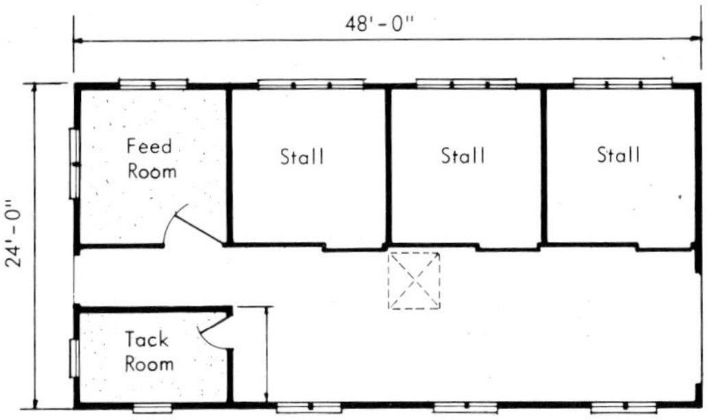

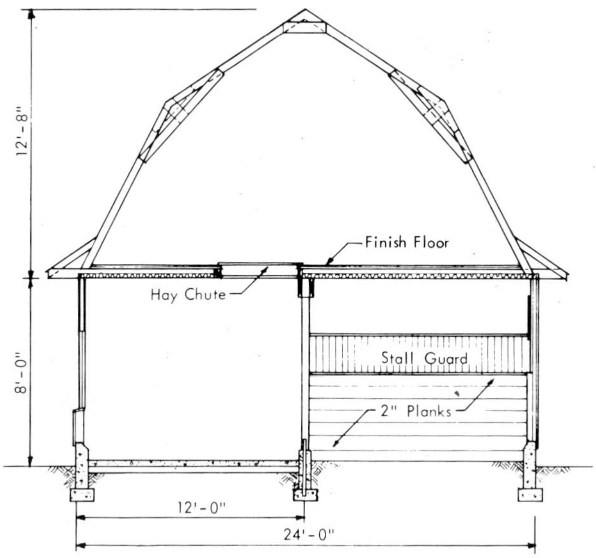

Four-Stall Barn USDA 5994

Comfortable quarters for four horses are provided by this 20' x 60' barn. Four 12' x 12' box stalls, as well as a combination tack and feed room, open to an 8' covered way.

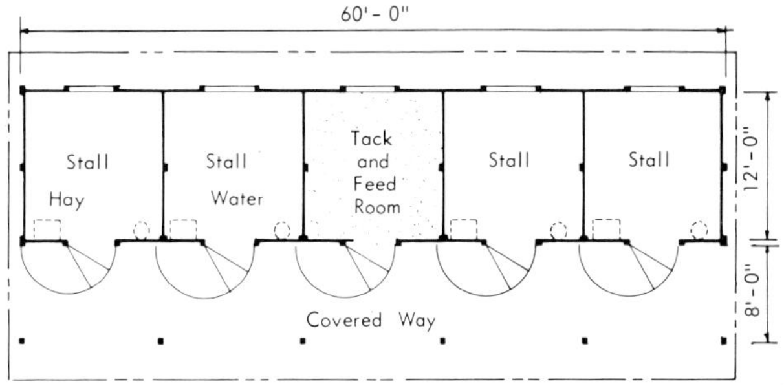

Eight-Stall Barn USDA 6010

Designed to give good protection in mild climates, this barn will house eight horses in 10' x 12' box stalls, placed back-to-back. Poles or posts are used as support members. The roof design provides for a 4' overhang on each side. The building can be enlarged in 10' increments.

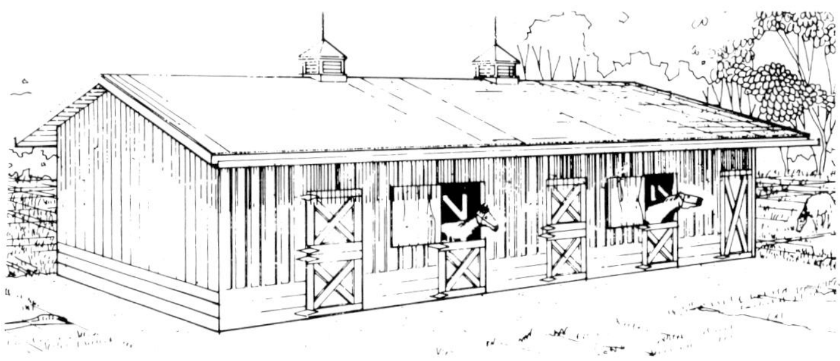

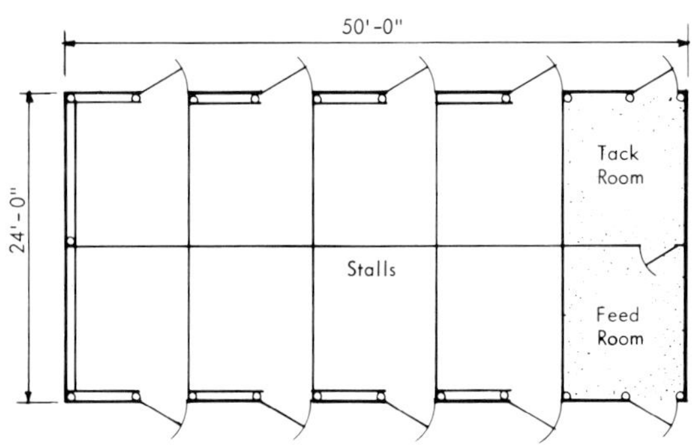

17-Stall Barn USDA 6011

This plan shows the largest barn in this list—36' x 100' with 17 box stalls, a 12' x 20' tack room, a 10' x 12' feed room and a 12' wide central service alley. Construction may include storage above a center alley supported on two rows of poles, or a clear-span trussed roof.

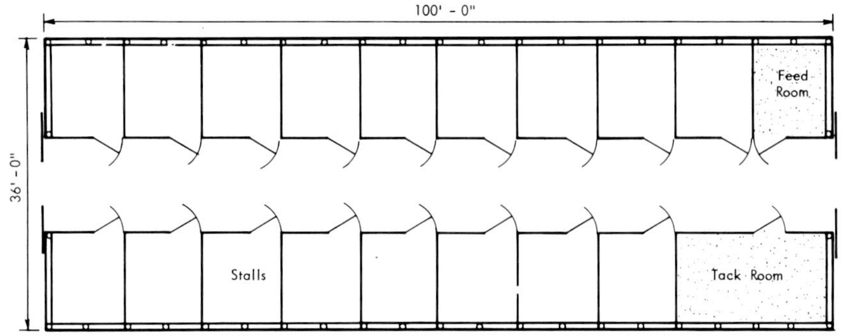

Horse Stalls USDA 5175

This plan can supplement the details in this book. Included are pictorial and detail drawings of face-in and face-out stalls that use feed alleys, and a wall-type stall that uses no feed alley.

Two-Horse Trailer USDA 5943

Complete construction details for a four-wheel, tandem axle trailer are shown on this plan. The frame is 1" pipe and 3" channel iron; there are no springs.

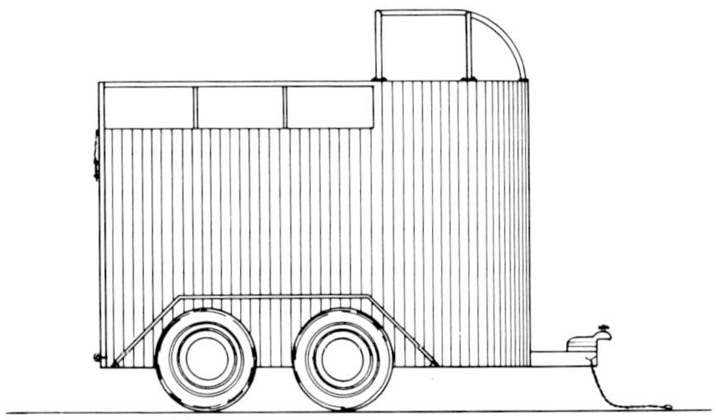

Other Publications of Interest

PLANS

Plan No.	Price	Title
Framing & Utility Buildings		
72008A	25¢	Braced Rafters For 32′ Barn Width
72008B	25¢	〃 〃 〃 34′ 〃 〃
72008C	25¢	〃 〃 〃 36′ 〃 〃
72009	25¢	Shawver Roof Truss
72010	25¢	Scissors Truss
72015	$1.00	Laminated Arches
72044	75¢	24′ Wide Frame Utility Building
72045	75¢	30′ 〃 〃 〃 〃
72046	$1.00	36′ 〃 〃 〃 〃
72047	75¢	40′ 〃 〃 〃 〃
72050	50¢	Pole Type Utility Building
72054	75¢	24′ Wide Pole Utility Building
72055	75¢	30″ 〃 〃 〃 〃
72056	$1.00	36′ 〃 〃 〃 〃
72057	75¢	40′ 〃 〃 〃 〃
72058	$1.00	50′ 〃 〃 〃 〃
72064	75¢	24′ Wide Rigid Frame Utility Building
72065	75¢	30′ 〃 〃 〃 〃 〃
72067	75¢	40′ 〃 〃 〃 〃 〃
72074	75¢	24′ Wide Masonry Utility Building
72075	75¢	30′ 〃 〃 〃 〃
72077	75¢	40′ 〃 〃 〃 〃
72084	75¢	24′ Wide Tilt-Up Concrete Utility Building
72085	75¢	30′ 〃 〃 〃 〃 〃
72087	75¢	40′ 〃 〃 〃 〃 〃
Hay Storage		
73103	50¢	50 Ton Self-Feeding Haykeeper
73104	50¢	24′ Wide Pole Hay Shed
73105	50¢	30′ 〃 〃 〃 〃
73106	50¢	36′ 〃 〃 〃 〃
73107	50¢	40′ 〃 〃 〃 〃
73108	50¢	48′ 〃 〃 〃 〃
Crop Storage		
73217	$1.00	Small Feed Bins
73250	$1.00	Small Storage Buildings
73271	25¢	Single Pole Corn Crib
73272	25¢	Single Wood Frame Corn Crib
73280	25¢	Double Pole Corn Crib
Machinery Storage		
74101	25¢	One-Car Garage
74102	25¢	Two-Car Garage
74119	$1.00	30′ x 30′ Farm Shop
74142	$1.00	Machinery Shed, Side Opening, 30′ Wide
74143	$1.00	Machinery Shed, End Opening, 40′ Wide

DIGESTS

Free—1 to 9 copies (total order)

AED–6	Gluing Wood
AED–9	Paints and Painting
AED–13	Insulation and Heat Loss

BOOKS

$1.00 each

MWPS–6	Beef Housing and Equipment Handbook
MWPS–8	Swine Housing and Equipment
MWPS–9	Designs for Glued Trusses
MWPS–10	Designs for Split Ring Trusses
MWPS–11	Designs for Nailed Trusses
MWPS–12	Outdoor Living Planning and Construction
MWPS–13	Planning Grain-Feed Handling
MWPS–14	Private Water Systems

$2.00 Each

MWPS—7	Dairy Housing and Equipment Handbook
MWPS–16	Family Housing Handbook

REFERENCE DATA

Table 15. Storage space requirements for feed and bedding

CROP	Lbs/Bu	Lbs/Cu Ft	Freshly Ground Lbs/Bu	Freshly Ground Lbs/Cu Ft	HAY-STRAW		Cu Ft/Ton	Lb/Cu Ft
Corn 15½%							LOOSE	
Shelled	56	44.8	48	38	Alfalfa		450-500	4.4-4
Ear	70	28.0	45	36	Non legume		450-600	4.4-3.3
Corn 30%	(Amounts to yield a bu or c ft of 15½% grain.)				Straw		670-1000	3-2
Shelled	67.5	54.0						
Ear, ground			89.6	35.8			BALED	
Barley 15%	48	38.4	37	28	Alfalfa		200-330	10-6
Flax 11%	56	44.8			Non legume		250-330	8-6
Grain Sorghum 15%	56	44.8			Straw		400-500	5-4
Oats 16%	32	25.6	23	18		CUT	CHOPPED	
Rye 16%	56	44.8	48	38	Alfalfa	1½"	285-360	7-5.5
Soybeans 14%	60	48.0			Non legume	3 "	300-400	6.7-5
Wheat 14%	60	48.0	50	43	Straw		250-350	8-5.7

Table 16. Light horse rations[1]

Age, Sex and Use	Live Weight lb	Live Weight kg	Daily Allowance (lb per 100 lb live weight) Grain	Hay	Ration Type[2]
Stallions in breeding season	900-1400	408-635	3/4-1 1/2	3/4-1 1/2	1, 2, 3
Pregnant mares	900-1400	408-635	3/4-1 1/2	3/4-1 1/2	4, 5, 6
Foals before weaning	100-350	45-159	1/2-3/4	1/2-3/4	4, 7, 8
Weanlings	350-450	159-204	1-1 1/2	1 1/2-2	8, 9, 10
Yearlings	450-700	204-318	–Summer Pasture–		
in training	450-700	204-318	3/4-1 1/4	3/4-1 1/4	3, 4, 8, 10
Yearlings, or rising 2 year olds	700-1000	318-454	1/2-1	1-1 1/2	3, 4, 11
Light horses at work					
hard use	900-1400	408-635	1/4-1 1/2	1-1 1/4	3, 12, 13
medium use	900-1400	408-635	3/4-1	1-1 1/4	3, 12, 13
light use	900-1400	408-635	1/2	1-1 1/4	3, 12, 13
Mature idle horses	900-1400	408-635	–	1 1/2-1 3/4	–

[1]From Light Horses, USDA Farmer's Bulletin 2127.
[2]Refer to Table 17.

Table 17. Grain rations for light horses[1]

Feed Material	1	2	3	4	5	6	7	8	9	10	11	12	13
	lb of Feed Material per 100 lb of Ration												
Barley	–	–	–	–	45	–	–	30	–	–	35	–	30
Corn	–	35	–	–	–	–	–	–	–	–	–	70	–
Oats	55	35	100	80	45	95	50	30	70	80	35	30	70
Wheat	20	15	–	–	–	–	–	30	–	–	–	–	–
Wheat bran	20	15	–	20	10	–	40	–	15	–	15	–	–
Linseed meal	5	–	–	–	–	5	10	10	15	20	15	–	–

[1]From Light Horses, USDA Farmer's Bulletin 2127.

Table 18. Water requirements

Type and age of animal	Gallons per head per day
Mature (mare or gelding)	8-12
Brood mares	8-12
Foals to 2-yr olds	6- 8
Stallions	8-12
Ponies	6- 8

CONVERSIONS

Multiply to the right. (Cu. Ft. x 7.5 = Gal.)
Divide to the left. (Gal. ÷ 7.5 = Cu. Ft.)

CUBIC FEET	7.5	Gallons
	0.4	Bu. Ear Corn
	0.8	Bu. Grain
	1728	Cubic Inches
GALLONS	231	Cubic Inches
	0.133	Cubic Feet
	8.3	Pounds Water
BUSHELS	1.25	Cubic Feet
	2.5	Cu. Ft. Ear Corn
CUBIC YARDS	27	Cubic Feet
concrete	81	Sq. Ft. 4" Floor
concrete	54	Sq. Ft. 6" Floor
ACRES	43,560	Square Feet
	4,840	Square Yards
	160	Square Rods
	1/640	Square Mile
MILES	5,280	Feet
	1,760	Yards
	320	Rods
RODS	16.5	Feet
	5.5	Yards

AREAS & VOLUMES

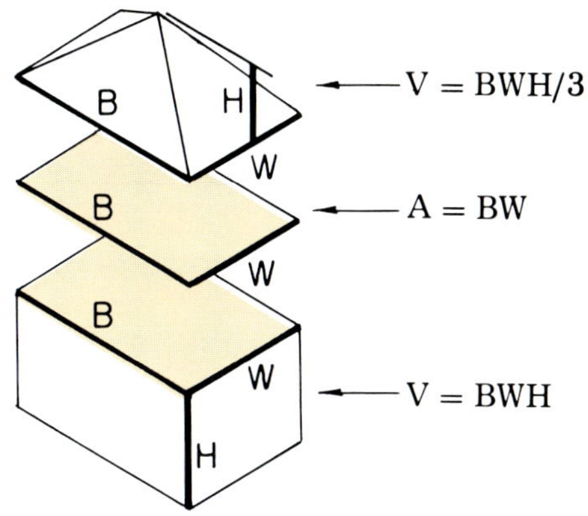

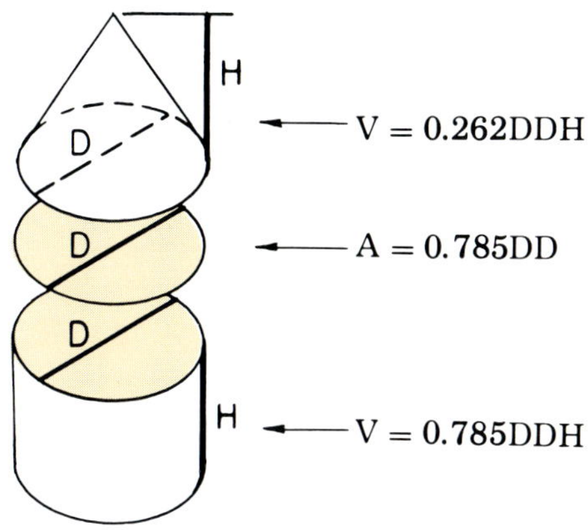

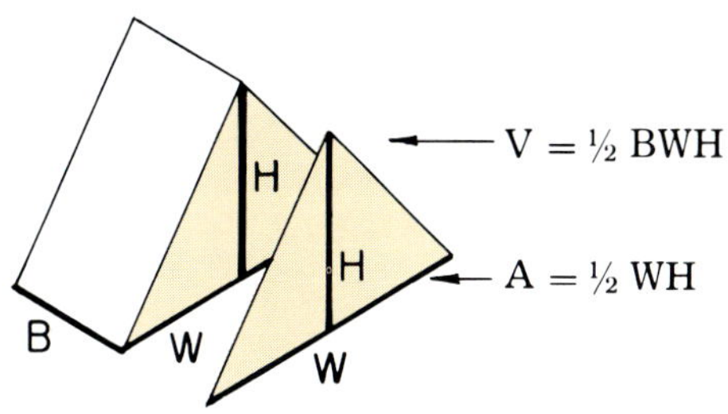

LUMBER

Board Foot = 1" x 1' x 1' nominal dimensions.
Board Feet = thickness in inches x width in feet (4" = 1/3') x length in feet.

i.e. : 1 x 4 x 10' = 1 x 1/3 x 10 = 3 1/3 fbm
2 x 6 x 12' = 2 x 1/2 x 12 = 12 fbm

Actual Dimensions: (Dressed 4 sides - S4S)

Nominal	1"	2"	3"	4"	6"	8"	10" etc
Actual	¾	1½	2½	3½	5½	7¼	9¼

1 x 6 measures ¾" x 5½"
2 x 8 measures 1½" x 7¼"